ZIML Math Competition Book

Junior Varsity 2017-2018

Areteem Institute

ZIML Math Competition Book Junior Varsity 2017-18

Edited by John Lensmire
David Reynoso
Kelly Ren
Kevin Wang

WWW.ARETEEM.ORG

PUBLISHED BY ARETEEM PRESS

ISBN: 1-944863-30-3
ISBN-13: 978-1-944863-30-2

First printing, November 2018.

TITLES PUBLISHED BY ARETEEM PRESS

Cracking the High School Math Competitions (and Solutions Manual) - Covering AMC 10 & 12, ARML, and ZIML
Mathematical Wisdom in Everyday Life (and Solutions Manual) - From Common Core to Math Competitions
Geometry Problem Solving for Middle School (and Solutions Manual) - From Common Core to Math Competitions
Fun Math Problem Solving For Elementary School (and Solutions Manual)

ZIML MATH COMPETITION BOOK SERIES

ZIML Math Competition Book Division E 2016-2017
ZIML Math Competition Book Division M 2016-2017
ZIML Math Competition Book Division H 2016-2017
ZIML Math Competition Book Jr Varsity 2016-2017
ZIML Math Competition Book Varsity Division 2016-2017
ZIML Math Competition Book Division E 2017-2018
ZIML Math Competition Book Division M 2017-2018
ZIML Math Competition Book Division H 2017-2018
ZIML Math Competition Book Jr Varsity 2017-2018
ZIML Math Competition Book Varsity Division 2017-2018
ZIML Math Competition Book Division E 2018-2019
ZIML Math Competition Book Division M 2018-2019
ZIML Math Competition Book Division H 2018-2019
ZIML Math Competition Book Jr Varsity 2018-2019
ZIML Math Competition Book Varsity Division 2018-2019

MATH CHALLENGE CURRICULUM TEXTBOOKS SERIES

Math Challenge I-A Pre-Algebra and Word Problems
Math Challenge I-B Pre-Algebra and Word Problems
Math Challenge I-C Algebra
Math Challenge II-A Algebra
Math Challenge II-B Algebra
Math Challenge III Algebra

Math Challenge I-A Geometry
Math Challenge I-B Geometry
Math Challenge I-C Topics in Algebra
Math Challenge II-A Geometry
Math Challenge II-B Geometry
Math Challenge III Geometry
Math Challenge I-A Counting and Probability
Math Challenge I-B Counting and Probability
Math Challenge I-C Geometry
Math Challenge II-A Combinatorics
Math Challenge II-B Combinatorics
Math Challenge III Combinatorics
Math Challenge I-A Number Theory
Math Challenge I-B Number Theory
Math Challenge I-C Finite Math
Math Challenge II-A Number Theory
Math Challenge II-B Number Theory
Math Challenge III Number Theory

COMING SOON FROM ARETEEM PRESS

Fun Math Problem Solving For Elementary School Vol. 2 (and Solutions Manual)
Counting & Probability for Middle School (and Solutions Manual)
- From Common Core to Math Competitions
Number Theory Problem Solving for Middle School (and Solutions Manual) - From Common Core to Math Competitions

The books are available in paperback and eBook formats (including Kindle and other formats).
To order the books, visit `https://areteem.org/bookstore`.

Contents

Introduction

Each month during the school year, Areteem Institute hosts the online Zoom International Math League (ZIML) competitions. Students can compete in one of five divisions based on their age and mathematical level (details shown on Page 9).

This book contains the problems, answers, and full solutions from the nine ZIML Junior Varsity Competitions held during the 2017-2018 School Year. It is divided into three parts:

1. The complete Junior Varsity ZIML Competitions (20 questions per competition) from October 2017 to June 2018.
2. The solutions for each of the competitions, including detailed work and helpful tricks.
3. An appendix including the topics and knowledge points covered for Junior Varsity, a glossary including common mathematical terms, and answer keys for each of the competitions so students can easily check their work.

The questions found on the ZIML competitions are meant to test your problem solving skills and train you to apply the knowledge you know to many different applications. We hope you enjoy the problems!

About Zoom International Math League

The Zoom International Math League (ZIML) has a simple goal: provide a platform for students to build and share their passion for math and other STEM fields with students from around the globe. Started in 2008 as the Southern California Mathematical Olympiad, ZIML has a rich history of past participants who have advanced to top tier colleges and prestigious math competitions, including American Math Competitions, MATHCOUNTS, and the International Math Olympaid.

The ZIML Core Online Programs, most available with a free account at ziml.areteem.org, include:

- **Daily Magic Spells:** Provides a problem a day (Monday through Friday) for students to practice, with full solutions available the next day.
- **Weekly Brain Potions:** Provides one problem per week posted in the online discussion forum at ziml.areteem.org. Usually the problem does not have a simple answer, and students can join the discussion to share their thoughts regarding the scenarios described in the problem, explore the math concepts behind the problem, give solutions, and also ask further questions.
- **Monthly Contests:** The ZIML Monthly Contests are held the first weekend of each month during the school year (October through June). Students can compete in one of 5 divisions to test their knowledge and determine their strengths and weaknesses, with winners announced after the competition.
- **Math Competition Practice:** The Practice page contains sample ZIML contests and an archive of AMC-series tests for online practice. The practices simulate the real contest environment with time-limits of the contests automatically controlled by the server.
- **Online Discussion Forum:** The Online Discussion Forum

is open for any comments and questions. Other discussions, such as hard Daily Magic Spells or the Weekly Brain Potions are also posted here.

These programs encourage students to participate consistently, so they can track their progress and improvement each year.

In addition to the online programs, ZIML also hosts onsite Local Tournaments and Workshops in various locations in the United States. Each summer, there are onsite ZIML Competitions at held at Areteem Summer Programs, including the International ZIML Convention, which is a two day convention with one day of workshops and one day of competition.

ZIML Monthly Contests are organized into five divisions ranging from upper elementary school to advanced material based on high school math.

- **Varsity:** This is the top division. It covers material on the level of the last 10 questions on the AMC 12 and AIME level. This division is open to all age levels.
- **Junior Varsity:** This is the second highest competition division. It covers material at the AMC 10/12 level and State and National MathCounts level. This division is open to all age levels.
- **Division H:** This division focuses on material from a standard high school curriculum. It covers topics up to and including pre-calculus. This division will serve as excellent practice for students preparing for the math portions of the SAT or ACT. This division is open to all age levels.
- **Division M:** This division focuses on problem solving using math concepts from a standard middle school math curriculum. It covers material at the level of AMC 8 and School or Chapter MathCounts. This division is open to all students who have not started grade 9.

- **Division E:** This division focuses on advanced problem solving with mathematical concepts from upper elementary school. It covers material at a level comparable to MOEMS Division E. This division is open to all students who have not started grade 6.

This problem book features the Junior Varsity Contests. For a detailed list of topics covered for Junior Varsity see p.159 in the Appendix.

About Areteem Institute

Areteem Institute is an educational institution that develops and provides in-depth and advanced math and science programs for K-12 (Elementary School, Middle School, and High School) students and teachers. Areteem programs are accredited supplementary programs by the Western Association of Schools and Colleges (WASC). Students may attend the Areteem Institute through these options:

- Live and real-time face-to-face online classes with audio, video, interactive online whiteboard, and text chatting capabilities;
- Self-paced classes by watching the recordings of the live classes;
- Short video courses for trending math, science, technology, engineering, English, and social studies topics;
- Summer Intensive Camps on prestigious university campuses and Winter Boot Camps;
- Practice with selected daily problems for free, and monthly ZIML competitions at ziml.areteem.org.

The Areteem courses are designed and developed by educational experts and industry professionals to bring real world applications into STEM education. The programs are ideal for students who wish to build their mathematical strength in order to excel academically and eventually win in Math Competitions (AMC, AIME, USAMO, IMO, ARML, MathCounts, Math Olympiad, ZIML, and other math leagues and tournaments, etc.), Science Fairs (County Science Fairs, State Science Fairs, national programs like Intel Science and Engineering Fair, etc.) and Science Olympiad, or purely want to enrich their academic lives by taking more challenges and developing outstanding analytical, logical thinking and creative problem solving skills.

Since 2004 Areteem Institute has been teaching with methodology that is highly promoted by the new Common Core State Standards: stressing the conceptual level understanding of the math concepts, problem solving techniques, and solving problems with real world applications. With the guidance from experienced and passionate professors, students are motivated to explore concepts deeper by identifying an interesting problem, researching it, analyzing it, and using a critical thinking approach to come up with multiple solutions.

Thousands of math students who have been trained at Areteem achieved top honors and earned top awards in major national and international math competitions, including Gold Medalists in the International Math Olympiad (IMO), top winners and qualifiers at the USA Math Olympiad (USAMO/JMO), and AIME, top winners at the Zoom International Math League (ZIML), and top winners at the MathCounts National. Many Areteem Alumni have graduated from high school and gone on to enter their dream colleges such as MIT, Cal Tech, Harvard, Stanford, Yale, Princeton, U Penn, Harvey Mudd College, UC Berkeley, UCLA, etc. Those who have graduated from colleges are now playing important roles in their fields of endeavor.

Further information about Areteem Institute, as well as updates and errata of this book, can be found online at `http://www.areteem.org`.

Acknowledgments

This book contains the Online ZIML Junior Varsity Problems from the 2017-18 school year. These problems were created and compiled by the staff of Areteem Institute. These problems were inspired by questions from the Areteem Math Challenge Courses, past questions on the ACT/SAT/GRE, past math competitions, math textbooks, and countless other resources and people encountered by the Areteem Curriculum Department in their life devoted to math. We thank all these sources for growing and nurturing our passion for math.

The Areteem staff, including John Lensmire, David Reynoso, Kevin Wang, and Kelly Ren, are the main contributors who compiled, edited, and reviewed this book.

Lastly, thanks to all the students who have participated and continue to participate in the Zoom International Math League. Your dedication to the Daily Magic Spells and Monthly Contests makes all of this possible, and we hope you continue to enjoy ZIML for years to come!

1. ZIML Contests

This part of the book contains the Junior Varsity ZIML Contests from the 2017-18 School Year. There were nine monthly competitions, held on the dates found below:

- October 6-8
- November 3-5
- December 1-3
- January 5-7
- February 2-4
- March 2-4
- April 6-8
- May 4-6
- June 1-3

1.1 ZIML October 2017 Junior Varsity

Below are the 20 Problems from the Junior Varsity ZIML Competition held in October 2017.
The answer key is available on p.173 in the Appendix.
Full solutions to these questions are available starting on p.74.

Problem 1

5 boys and 4 girls compete in a race. If none of the girls finish one after another (second and third, eighth and ninth, etc.), how many possible outcomes are there?

Problem 2

Peter and John each have a fair coin. Peter flips the coin 3 times while John flips the coin 5 times. The probability that Peter and John get the same number of tails can be written as $\frac{P}{Q}$ with P, Q relatively prime positive integers. What is $P+Q$?

Problem 3

$\triangle ABC$ has sides $AB = AC = 12$ and $BC = 4$. Point D is on AB and point E is on AC such that $\overline{DE}$ is parallel to $\overline{BC}$. If the perimeter of ADE is the same as $DECB$, that is the length of AD?

Problem 4

Find the sum of all integer solutions to

$$x(x-1)(x^2+11x+30) = 72.$$

Problem 5

A five-digit odd number has five distinct digits, and it is divisible by 9. What is the largest such number?

Problem 6

Let $\triangle ABC$ have perimeter 72. Point E is on BC with $BE = 8$ and $CE = 10$. If $\angle EAB = \angle EAC$, what is the length of AC? Round your answer to the nearest integer if necessary.

Problem 7

Consider the number 2000. How many ways are there to represent it as the product of three factors? Here we consider $20 \cdot 10 \cdot 10$ to be different from $10 \cdot 20 \cdot 10$.

Problem 8

Find the sum of all positive n such that $x^2 - nx - 24$ factors.

Problem 9

$(20, 21, 29)$ is a triple with $a^2 + b^2 = c^2$ (called a Pythagorean triple). The sum of this triple is $20 + 21 + 29 = 70$. There is one other Pythagorean triple containing the number 29. What is the sum of this triple?

Problem 10

Calculate $2017^{2018} \pmod{20}$.

Problem 11

$ABCD$ is a square with side length 4. Construct equilateral triangle EFG with E, F, and G on the sides of square $ABCD$ so that the area of EFG is as small as possible. This area can be written as $\sqrt{K}$ for an integer K. What is K?

Problem 12

Two circles, C_1 and C_2, of radii 4 and 9 respectively, are tangent. Both circles are tangent to the line $y = 0$. The line $x = a$ is tangent to C_1 but does not intersect C_2 while the line $x = b$ is tangent to C_2 but does not intersect C_1. What is $|b - a|$? Round your answer to the nearest tenth if necessary.

Problem 13

In the diagram below, there are 21 grid points arranged in equilateral triangles, equally spaced.

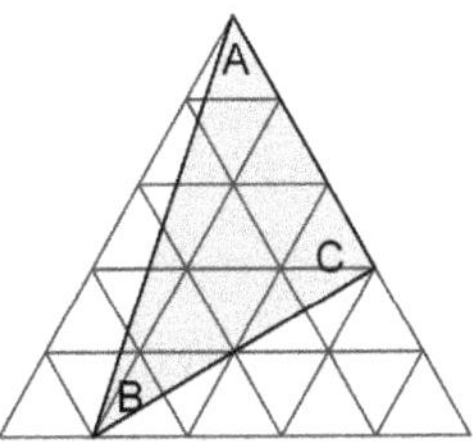

What percentage of the diagram is shaded? Give you answer as an integer rounded to the nearest percent if necessary.

Problem 14

Recall $\lfloor x \rfloor$ denotes the largest integer $\leq x$. Set $\{x\} = x - \lfloor x \rfloor$. The smallest solution to $\lfloor x \rfloor \cdot \{x\} = 4$ can be written as $\frac{p}{q}$ with $\gcd(p,q) = 1$. What is $p+q$?

Problem 15

Find the sum of all integers m such that the positive difference of the roots of $x^2 - mx - 8 = 0$ is $2 - m$.

Problem 16

Wanda puts 8 different rings on the 4 fingers (not her thumb) of her left hand. If she puts 2 rings on each finger, how many different arrangements of these rings are there?

Problem 17

Suppose you have a fair 5-sided die (numbered $1-5$). You roll it 3 times. How many of these outcomes have an odd sum?

Problem 18

How many factors of $6,250,000$ are perfect squares? Remember 1 and $6,250,000$ are factors of $6,250,000$.

Problem 19

How many times does the graph $y = |x^2 - 2x - 3| - |x^2 + x - 2|$ intersect the x-axis?

Problem 20

There is one positive integer a such that the equation

$$x^2 + (a - 10)x + a = 0$$

has two integer roots. What is this a?

1.2 ZIML November 2017 Junior Varsity

Below are the 20 Problems from the Junior Varsity ZIML Competition held in November 2017.
The answer key is available on p.174 in the Appendix.
Full solutions to these questions are available starting on p.81.

Problem 1

Find the smallest 3-digit number n such that the product of the divisors of n is equal to n^2.

Problem 2

Let's consider the roots, multiplicities included, to the equation $(2x^2+9x+4)^3=0$. What is the product of all the roots?

Problem 3

The equation $\frac{1}{\sqrt{x+2}}+\sqrt{x+2}=\frac{10}{3}$ has one solution of the form $\frac{P}{Q}$ for P,Q integers, $Q>1$, and $\gcd(P,Q)=1$. What is $P-Q$?

Problem 4

How many numbers between 2017 and 7102 are divisible by 20 but NOT divisible by 17?

Problem 5

Suppose you have the numbers 1, 2, 3, 4, 5, and 6. How many different 6-digit numbers can be formed with the digit 2 next to either the digit 1 or the digit 3?

Problem 6

Let $ABCD$ be a parallelogram with area 40. Suppose P is a point inside the parallelogram such that the area of $\triangle ABP = 7$ and the area of $\triangle BCP = 9$. What is the area of $\triangle CDP$, rounded to the nearest integer if necessary?

Problem 7

A 5-digit number, all of whose digits are distinct even numbers, is divisible by 11. What is the smallest such number?

Problem 8

$\triangle ABC$ has side lengths of 20, 13, and 11. Consider the 3 altitudes of the triangles. What is the length of the shortest altitude, rounded to the nearest integer if necessary?

Problem 9

Suppose $PENTA$ is a regular pentagon. Equilateral triangle TAB is drawn sharing side $\overline{AT}$ with the pentagon, so that $\angle BAP$ is as large as possible (with $\angle BAP < 180^\circ$). What is the measure of $\angle BAP$ in degrees?

Problem 10

There is one pair of nonzero real numbers x and y satisfying $x+|y|=2$ and $x|y|-3x^3=0$. Find the value of y in decimal, rounded to the nearest hundredth if necessary.

Problem 11

Suppose you have 8 red cards and 16 black cards. Assume all the cards of the same color are identical. Deal the cards out in a line. How many arrangements of the cards are there if there must be at least 2 black cards between any two of the red cards?

Problem 12

Suppose you randomly pick a real number on the number line between -5 and 15. The probability that the number squared is greater than 16 can be expressed as $N\%$. What is N, rounded to the nearest integer?

Problem 13

The equation $x^2+(a-6)x+a=0$ has two integer roots. If $a>0$, find the value of a.

Problem 14

Suppose chords $\overline{AB}, \overline{CD}$ intersect at E, such that $AE:EB=1:4$ and $CE:ED=4:9$. The ratio of $AB:CD$ can be written as a fraction $\frac{P}{Q}$ for positive integers P, Q with $\gcd(P,Q)=1$. What is $P+Q$?

Problem 15

Suppose you flip a fair coin. If you get heads, you roll a fair 6-sided die twice. If you get tails, you roll a fair 6-sided die three times. The probability that the sum of the rolls is 8 can be written as $\frac{P}{Q}$ for positive integers P, Q with $\gcd(P,Q) = 1$. What is $Q - P$?

Problem 16

Suppose you write out the numbers $1-2017$: $1,2,3,4,\ldots,2017$. How many digits have you written in total?

Problem 17

Find $\gcd(2^{38}-1, 2^{26}-1)$.

Problem 18

What is the largest integer K such that $\sqrt{2x} - \sqrt{2x-10} = K$ has at least one real solution?

Problem 19

Find the remainder of 3^{3^3} when it is divided by 11.

Problem 20

Trapezoid $ABCD$ with $\overline{AB}$ parallel to $\overline{CD}$ is inscribed in a circle of radius 4. What is the area of $ABCD$ if $BC = CD = 4$? Round your answer to the nearest integer.

1.3 ZIML December 2017 Junior Varsity

Below are the 20 Problems from the Junior Varsity ZIML Competition held in December 2017.
The answer key is available on p.175 in the Appendix.
Full solutions to these questions are available starting on p.90.

Problem 1

Suppose you have a circle with diameter $\overline{AB}$ with $AB = 12$. Let C, D be on arc $\widehat{AB}$ such that $\widehat{AC} : \widehat{CD} : \widehat{DB} = 1 : 4 : 1$ (ratio of the arc lengths). The area of the figure enclosed by line segment $\overline{AC}$, arc $\widehat{CD}$, and line segment $\overline{AD}$ can be written as $K \times \pi$ for an integer K. What is K?

Problem 2

Consider five-digit numbers of the form $\overline{a357b}$. What is the sum of all such numbers that are divisible by 36?

Problem 3

Suppose a man has 4 sons and 3 daughters, and has 3 boys' schools and 4 girls' schools available to choose from. How many different ways can he send his children to school so none of his daughters attend the same school?

Problem 4

How many solutions are there to $\frac{1}{x} - \frac{1}{y} = \frac{1}{3}$ if x, y are allowed to be any integers (positive or negative)?

Problem 5

Let p, q, and r be prime numbers, such that the product pqr is 19 times the sum $p+q+r$. Find the sum of all possible values of $p^2+q^2+r^2$.

Problem 6

Find the smallest positive integer that leaves remainder 1 when divided by 3, remainder 0 when divided by 5, remainder 2 when divided by 7, and 4 when divided by 9.

Problem 7

Solve for real x: $\sqrt{\dfrac{x-2}{x+2}}+\sqrt{\dfrac{9x+18}{x-2}}=4$. The sum of all the real solutions can be written as $\dfrac{P}{Q}$ for integers P,Q with $\gcd(P,Q)=1$. What is $P\cdot Q$?

Problem 8

A bag contains 5 red, 4 green, and 3 yellow balls. You pick 3 balls at once (so in no particular order). The probability you get all 3 balls of the same color can be written as $\dfrac{P}{Q}$ for positive integers P,Q with $\gcd(P,Q)=1$. What is $Q-P$?

Problem 9

Draw a circle tangent to the lines $y=x$, $y=-x$, and $x=4$. The radius of this circle can be expressed as $R+\sqrt{S}$ for integers R,S. What is $R+S$?

Problem 10

A certain carnival game is played as follows: a dart is dropped onto a circular board with radius 4 inches. To win the game the dart must be dropped into a smaller circular region with radius 2 inches. (Assume the dart always hits the board, but in a random place.) The game starts by flipping a (fair coin), which gives you one or two tries (heads is one try, tails is two tries). The probability of winning is approximately $W\%$ where W is an integer. What is W?

Problem 11

Find the smallest solution to $\lfloor x \rfloor + 4\{x\} = 8$. Round your answer to the nearest hundredth if necessary. Here $\lfloor x \rfloor$ denotes the largest integer $\leq x$ and $\{x\} = x - \lfloor x \rfloor$.

Problem 12

A shop has 8 types of cookies and you want to buy a total of 6 cookies. In how many ways can you buy the cookies if you want to make sure you have an even number of each type of cookie you buy? For example, you could buy 4 chocolate chip and 2 sugar cookies, or 6 sugar cookies, but not 5 chocolate chip and 1 sugar cookie.

Problem 13

The equation $4(x+1)^4 + 7x(x+1)^2 = 2x^2$ has one pair of real solutions that can be written in the form $A \pm \sqrt{B}$ for integers A, B. What is $A - B$?

Problem 14

Consider the sequence a_n defined by $a_1 = a_2 = 1$ and

$$a_n = 2a_{n-1} + a_{n-2}.$$

Find the remainder when a_{2018} is divided by 4.

Problem 15

Let x be a real number such that $x^3 + 9x = 27$. Determine the value of $x^7 + 486x^2$.

Problem 16

Let O be the center of a circle with radius 4. Points A and B are on the circle with $\angle AOB = 30^\circ$. Draw a line L_1 containing A parallel to $\overline{OB}$ and a line L_2 containing B parallel to $\overline{OA}$. Call the intersection of L_1 with the circle point C and the intersection of L_1 with L_2 point D. What is the area of trapezoid $BOCD$? Round your answer to the nearest integer.

Problem 17

Form a triangular pyramid with a base that is an equilateral triangle with side length 2. If the three sides of the pyramid are isosceles triangles with sides $4, 4, 2$, the volume of the pyramid can be expressed as $\frac{1}{3} \times \sqrt{K}$ for an integer K. What is K?

Problem 18

For how many integers m is it true the equation

$$x^2 + 2mx + 12 - m = 0$$

has two real roots (or one double root) that are each greater than or equal to 1?

Problem 19

Let $ABCD$ be a square and let P, Q be on sides $\overline{AD}, \overline{AB}$ respectively such that $DP = 1$, $BQ = 2$, and $PQ = 3$. Find the measure of $\angle PCQ$ in degrees.

Problem 20

Suppose you give out the 5 (currently published) "A Song of Ice and Fire" books to 3 of your friends. Each friend gets at least one book. How many ways an you give out the books?

1.4 ZIML January 2018 Junior Varsity

Below are the 20 Problems from the Junior Varsity ZIML Competition held in January 2018.
The answer key is available on p.176 in the Appendix.
Full solutions to these questions are available starting on p.100.

Problem 1

An 10-digit number $\overline{832xy1599}$ has remainder 9 when divided by 99. What is $10x+y$?

Problem 2

A spherical ball was floating on the lake when the lake was frozen. The ball was removed without breaking the ice. A hole was left on the surface of the ice, with diameter 20 and depth 6. The radius of the ball can be written as $\frac{P}{Q}$ for positive integers P,Q with $\gcd(P,Q)$. What is $P+Q$?

Problem 3

A two digit number equals the sum of its tens digit and the square of its units digit. What is this two digit number?

Problem 4

The diagonals of rectangle $ABCD$ intersect at a point E. Point E is 5 inches farther from $\overline{BC}$ than it is from $\overline{CD}$. If the area of the rectangle is 56 square inches, what is the perimeter in inches? Round your answer to the nearest integer if necessary.

Problem 5

What is the coefficient of x^4 in $(x^4+3)(x^2+3x+4)(x^2-2x-3)$?

Problem 6

Calculate the remainder when 9^{2018} is divided by 11.

Problem 7

Four couples, the Smiths, the Johnsons, the Garcias, and the Lees, meet up at their friends wedding. The four couples decide to all line up and take a photograph together. The two people in each couple stand next to each other, and also the Lees stand next to the Smiths. How many such photos are there?

Problem 8

Let a,b,c be the roots of $\frac{x^3}{5} - 10x^2 + 12x + 25 = 0$. Find the value of $\frac{1}{ab} + \frac{1}{ac} + \frac{1}{bc}$. Round your answer to the nearest tenth if necessary.

Problem 9

Let $ABCD$ be a parallelogram as in the diagram, with E the midpoint of $\overline{BC}$.

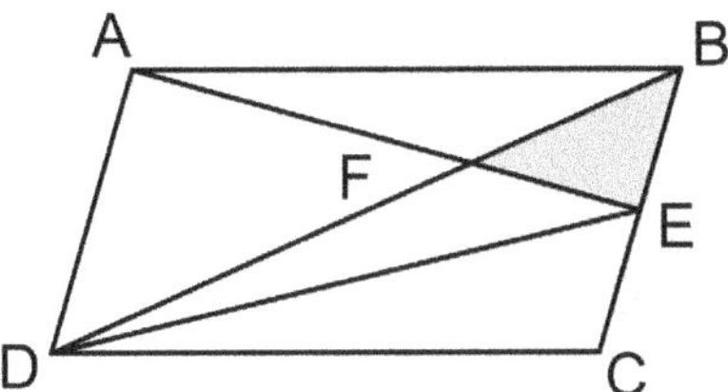

$\triangle BEF$ is $P\%$ of the entire parallelogram. What is P, rounded to the nearest tenth?

Problem 10

Consider subsets of $\{1,2,4,5,6,8,9,10\}$. How many subsets contain at least one odd and at least one even number?

Problem 11

Consider the first 100 perfect squares: $1^2, 2^2, \ldots, 100^2$. How many of these have remainder 1 when divided by 12?

Problem 12

Let O be the center of a circle with radius 4. Let $\widehat{AB}$ be an arc with measure 60° and C be the midpoint of $\overline{OB}$. Extend the line containing A and C, intersecting the circle again at point D. The length AD can be written as $\sqrt{L}$ for an integer L. What is L?

Problem 13

What is the product of all the integer solutions to the following equation?

$$(x^2-1)(x^2+10x+24)=24$$

Problem 14

Suppose you flip a fair coin 8 times. The probability you get more heads than tails can be written as $\frac{P}{Q}$ for positive integers P, Q with $\gcd(P, Q) = 1$. What is $P + Q$?

Problem 15

In the following diagram, $\triangle ABC$ has area 78 and G is the centroid of $\triangle ABC$.

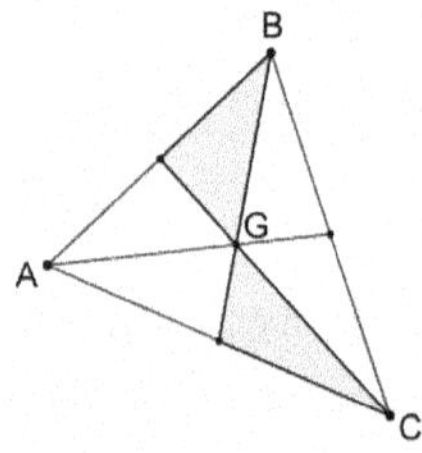

This shaded area is an integer K. What is K?

Problem 16

Let

$$f(x) = |x^2 - 2x - 3| - |x^2 + x - 2|.$$

Find the smallest L and largest H so that for all x satisfying $L \leq x \leq H$ we have $f(x) = 3x + 1$. What is $L + H$? Round your answer to the nearest integer if necessary.

Problem 17

Carrie invites 9 of her friends for dinner. Carrie will sit with four of the friends at the first circular table, while the other 5 friends sit around the second table. If all the seats at both tables are indistinguishable, how many seating arrangements are there?

Problem 18

Consider ordered pairs (x,y) of real numbers such that

$$x^2 - xy + y^2 = 13 \text{ and } x - xy + y = -5.$$

In fact, for all such pairs, $x + y$ is an integer. What is the largest such $x + y$?

Problem 19

You roll a fair 6-sided die once (numbered one through six). Then pick a real number at random in the interval $(0, K)$, where K was the number rolled. The probability that the real number is less than 2 can be expressed as $\frac{P}{Q}$ where P, Q are positive integers with $\gcd(P, Q) = 1$. What is $Q - P$?

Problem 20

There is one ordered triple (x,y,z) of prime numbers satisfying the equation $x(x+y)(y+z) = 140$ with $x \neq y, x \neq z, y \neq z$. For this triple, what is $x+y+z$?

1.5 ZIML February 2018 Junior Varsity

Below are the 20 Problems from the Junior Varsity ZIML Competition held in February 2018.
The answer key is available on p.177 in the Appendix.
Full solutions to these questions are available starting on p.108.

Problem 1

A family has 4 children, including at least one boy and at least one girl. Assume that every child was equally likely to be a boy or a girl. What is the probability that the family has at most two girls? Input your answer as $K\%$, where K is rounded to the nearest integer. (For example, if the probability is 49.2%, you would input 49 below.)

Problem 2

O is the center of a circle with radius 6. Let $ABCO$ be a rhombus with A, B, C on the circle. The area of $ABCO$ can be written in simplest radical form as $P\sqrt{Q}$ (P and Q are positive integers so that Q does not contain any perfect squares as factors). What is $P+Q$?

Problem 3

Consider pairs (x, y) satisfying the system of equations $|x|+y=2$ and $x+y^2=4$. What is the largest possible value of x^2+y^2? Round your answer to the nearest integer if necessary.

Problem 4

Find the remainder when $2016^{2017} + 2020^{2017}$ is divided by 2018.

Problem 5

Let a, b, c be the three roots of $x^3 - 4x + 3 = 0$. Calculate

$$a^3 + b^3 + c^3.$$

Round your answer to the nearest integer.

Problem 6

How many 4-digit numbers $\overline{abcd}$ (so $a, b, c, d \in \{0, 1, \ldots, 9\}$ with $a \neq 0$) are there such that the sum of $\overline{abc}$ and $\overline{bcd}$ is divisible by 11? (Here $\overline{abcd}$ denotes the 4-digit number with digits a, b, c, d.)

Problem 7

Inscribe a regular dodecagon inside a circle of radius 4. What is the area of the dodecagon, rounded to the nearest hundredth if necessary? (A dodecagon is a polygon with 12 sides.)

Problem 8

Pretend that everyone likes all of the five Maze Runner books equally, so every person has an equal chance of picking any of the books as their favorite. You poll 5 people and record their favorite book as one of the numbers $1,2,3,4,5$. The probability that the median of these numbers (the favorite books) is 5 (the last book) can be written as $\frac{P}{Q}$ for positive integers P,Q with $\gcd(P,Q)=1$. What is $P+Q$?

Problem 9

The range of $y=\sqrt{2x^2-12x+72}$ contains all the integers greater than or equal to K. What is the smallest integer K so that the preceding sentence is true?

Problem 10

The neighborhood pizza place has 5 toppings available. You want to order 2 different pizzas, each with 3 toppings. You do not order repeated toppings on a single pizza and the order of the toppings on a pizza does not matter. (However, it is possible to have a topping repeated on separate pizzas.)

If you only care which two pizzas you get, how many ways are there to make the order?

Problem 11

A cube $ABCD-EFGH$ with side length 2 is cut through $\triangle AFH$ into two smaller solids, one of which is a triangular pyramid. The surface area of the other solid created by this cut can be written as $A+B\sqrt{C}$ for integers A,B,C, where C does not have any perfect squares greater than 1 as factors. What is $A+B+C$?

Problem 12

From the set $\{1,2,\ldots,500\}$, select k numbers. What is the minimum value of k such that it is guaranteed to have two numbers with 5 as a common divisor?

Problem 13

Find the real solutions to $4x^2+4x-7-\frac{2}{x}+\frac{1}{x^2}=0$. What is the sum of all the irrational solutions? Round your answer to the nearest tenth if necessary.

Problem 14

Wendy flipped a coin 14 times, getting 4 heads and 10 tails. She remarked that she always got at least 2 tails between any two heads. How many different outcomes match Wendy's description?

Problem 15

Four identical balls (spheres), each of radius 1 in, are glued to the ground so that their centers form the vertices of a square with side length 2 in. Suppose you rest a fifth identical ball on the four balls (so the fifth ball is a sphere externally tangent to the other spheres). How far does this ball rest off the ground? Round your answer to the nearest tenth of an inch.

Problem 16

Suppose that A has 9 divisors and B has 4 divisors. Find $A+B$ if $\gcd(A,B)=7$ and $\text{lcm}(A,B)=2205$.

Problem 17

The equation

$$\left\lfloor \frac{2x-1}{3} \right\rfloor = \frac{3x-1}{2}$$

has one real solution. What is this solution, rounded to the nearest hundredth if necessary? Here $\lfloor a \rfloor$ denotes the largest integer not exceeding a.

Problem 18

Find the coefficient of x^{11} in $(x^3+x^2-1)^9$.

Problem 19

The equation $x^2+(a+14)x-a=0$ has two integer roots. Find the sum of all possible values of a.

Problem 20

A triangle has altitudes with ratio $12 : 15 : 20$ and area 96. The perimeter of the triangle can be written as $\sqrt{K}$ for an integer K. What is K?

1.6 ZIML March 2018 Junior Varsity

Below are the 20 Problems from the Junior Varsity ZIML Competition held in March 2018.
The answer key is available on p.178 in the Appendix.
Full solutions to these questions are available starting on p.116.

Problem 1

A five-digit number $\overline{abcde}$ has digits $5, 6, 7, 8, 9$ (not necessarily in that order). Assume that $5 \mid \overline{abcde}$, $4 \mid \overline{abcd}$, $3 \mid \overline{abc}$, and $2 \mid \overline{ab}$. Find the smallest possible value of $\overline{abcde}$. (For example, $3 \mid \overline{abc}$ means that 3 is a factor of the three-digit number with digits a, b, and c.)

Problem 2

Let E be the midpoint of side AB in square $ABCD$. Point F is the intersection of $\overline{BD}$ with $\overline{CE}$, point G the intersection of $\overline{AC}$ with $\overline{BD}$, and point H the intersection of $\overline{AC}$ with $\overline{DE}$. The area of quadrilateral $EFGH$ is $P\%$ of square $ABCD$. What is P, rounded to the nearest tenth if necessary? (For example, if $EFGH$ is $\frac{2}{3}$ of the entire square, we would have $P = 66.7$.)

Problem 3

10 friends decide to play five versus five basketball, so they need to divide themselves into two teams. If there is no order associated with how the people are picked and no order associated with the teams, how many different ways can they divide themselves into the two teams?

Problem 4

Consider the equation $x^3 - x^2 = y^2 - x^2y + xy - y$. What is the sum of all real numbers $a < 0$ such that $(x, y) = (a, 5)$ satisfies the equation? Round your answer to the nearest tenth if necessary.

Problem 5

A Poke restaurant offers 5 choices of seafood: tuna, albacore, salmon, shrimp, or octopus, served with either white or brown rice. A small Poke dish contains three portions of seafood with rice, while a large Poke dish contains four portions of seafood with rice. (For either size, repeated portions of seafood are allowed.) How many different Poke dishes are possible to order at this restaurant?

Problem 6

A quarter circle with radius 1 is drawn inside a unit square. A right triangle is drawn so that the legs are contained in the sides of the square and the hypotenuse is tangent to the circle as shown in the diagram below.

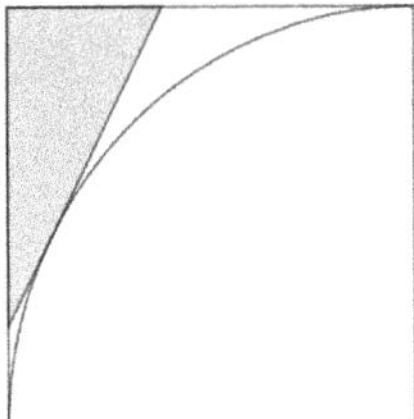

If the ratio of sides of the two legs is $1:2$, the length of the shorter leg can be written as $A+B\sqrt{C}$ where A and B are rational numbers and C is an integer (not containing any squares greater than 1 as factors). What is $A+B+C$, rounded to the nearest tenth if necessary?

Problem 7

Solve the equation $x^4+4x^3+2x^2-4x+5=4x^2+8x-4$. What is the smallest solution? Round your answer to the nearest tenth if necessary.

Problem 8

Consider a recursive sequence with $a_0=3$ and $a_{n+1}=9a_n-1$. What is the remainder when a_{2018} is divided by 80?

Problem 9

Let $x = \dfrac{1}{2+\sqrt{3}}$, $y = \dfrac{1}{11+\sqrt{12}}$. The expression

$$\frac{x^3y^3}{x^3 - 6x^2y + 12xy^2 - 8y^3}$$

can be written as a fraction $\dfrac{P}{Q}$ for positive P, Q with $\gcd(P, Q) = 1$. What is $P + Q$?

Problem 10

Justine rolls a fair 6-sided die three times, and arranges the rolls to form a 3-digit number. For example, if she rolls 1, 2, 2 she could form the numbers 122, 212 or 221. The probability that Justine can form a number that is divisible by 11 can be written as $\dfrac{P}{Q}$ where P and Q are positive integers with $\gcd(P, Q) = 1$. What is $P + Q$?

Problem 11

Consider the integers $1, 2, 3, \ldots, 100$. K of these integers have exactly L factors, where L is the maximum number of factors possible (so none of the integers $1, 2, \ldots, 100$ has more than L factors). What is $K \times L$?

Problem 12

Vance, Wally, Xander, Yondu, and Zeno run in a race and finish in that order (Vance in first, Wally in second, etc.). They have a rematch and Zeno remarks that no one finished in the same place as the the first race. How many such outcomes are possible?

Problem 13

Points A, B, and C are all on the same circle. Construct a line tangent to the circle at A and suppose the ray $\overrightarrow{BC}$ intersects this line at point D. Form trapezoid $BCEF$ with E and F on $\overleftrightarrow{AD}$, and $\overline{CE}$ and $\overline{BF}$ perpendicular to $\overleftrightarrow{AD}$. If $AD = 6\sqrt{5}$, $BC = 8$, and $CE = 6$, find the area of $BCEF$. Round your answer to the nearest integer if necessary.

Problem 14

Write $10! = A \cdot B \cdot C \cdot D$ for positive integers $A \leq B \leq C \leq D$. What is the smallest possible value of $D - A$?

Problem 15

Consider the cube shown below with a side length of 4.

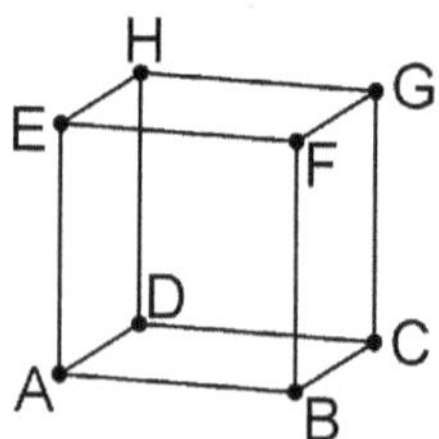

Let J be the center of face $ABFE$. Consider the shortest path from J to G that travels only in the faces $ABFE$, $ABCD$, and $CDHG$. What is the length of the portion of this path contained in face $ABCD$, rounded to the nearest integer?

Problem 16

The equation $x^2+ax+1=b$ has two integer roots. If $a^2+b^2=10$, how many different pairs (a,b) are there?

Problem 17

Kate and Leo play a card game. Their deck of cards consists of 25 cards, 5 with value 1, 5 with value 2, up to 5 with value 5. The game is simple, Kate picks a card, and then Leo picks a different card and whoever has the higher valued card wins. If they pick the same value, they tie. Suppose they play the game once and tie. They play again without shuffling the cards. The probability they tie again can be expressed as $\frac{P}{Q}$ for positive integers P, Q with $\gcd(P,Q)=1$. What is $Q-P$?

Problem 18

Find the sum of all values of a such that the equation

$$|x^2+ax+1| = |x^2+x+a|$$

has exactly two real solutions.

Problem 19

Circle C_1 with radius 1 is tangent to line segments $\overline{AB}$ and $\overline{AC}$ at B and C respectively so that $\angle BAC = 60^\circ$. Circle C_2 is tangent to C_1 at C and to the extension of ray $\overrightarrow{BA}$ at point D. What is the radius of C_2, rounded to the nearest tenth if necessary?

Problem 20

What is the sum of all integers K such that $\sqrt{K}$ is a real solution to the equation $x^2 = \lfloor 4x \rfloor$? (Here $\lfloor x \rfloor$ represents the greatest integer not exceeding x)

1.7 ZIML April 2018 Junior Varsity

Below are the 20 Problems from the Junior Varsity ZIML Competition held in April 2018.
The answer key is available on p.179 in the Appendix.
Full solutions to these questions are available starting on p.128.

Problem 1

Find the last two digits of 7^{7^7}.

Problem 2

Suppose you expand $(x-1)^7$ to get a polynomial (with degree 7). What is the sum of all the coefficients of this polynomial?

Problem 3

8 people get in an elevator on Floor 0. Each person leaves on one of Floor 1, Floor 2, etc., up to Floor 9 and the elevator only stops if someone leaves on that floor. How many different collections of floors can the elevator stop on?

Problem 4

Mark P inside square $ABCD$, so that triangle ABP is equilateral. Let Q be the intersection of BP with diagonal AC. Consider triangle $\triangle CPQ$. What is the largest angle in $\triangle CPQ$ minus the smallest angle in $\triangle CPQ$? Round your answer to the nearest degree if necessary.

Problem 5

Find the sum of all 5-digit integers, formed only using the digits 3 and 5, that are divisible by 15.

Problem 6

Consider the expression

$$L = \frac{2018\sqrt{2018}(\sqrt{2018}+4) - 3\sqrt{2018}(\sqrt{2018}-4)+1}{2018+4\sqrt{2018}}.$$

Calculate $\lfloor L \rfloor$. (Here $\lfloor x \rfloor$ is the greatest integer not exceeding x.)

Problem 7

Consider a unit cube. How many different planes pass through exactly 4 vertices of the cube?

Problem 8

Consider the expression $3+6+9+12+\cdots+42$. Suppose you can erase some, all, or none of the numbers. How many possible sums are there? (If you erase all the numbers, the sum is 0).

Problem 9

From the set of integers greater than or equal to 2 $(2,3,4,\ldots)$, delete 10 subsets as follows. First delete all even numbers except 2, then all multiples of 3, except 3, then all multiples of 5, except 5, and so on for the ten primes 2, 3, 5, 7, 11, 13, 17, 19, 23, and 29. After these steps, what is the smallest number remaining with 4 or more factors?

Problem 10

Suppose $ABCD$ is a convex quadrilateral whose diagonals $\overline{AC}$ and $\overline{BD}$ intersect at M. Suppose $BD = 8$ and $\triangle ABC$ and $\triangle ADC$ have areas 10 and 6 respectively. What is the length of BM? Round your answer to the nearest tenth if necessary.

Problem 11

George was watching his favorite basketball player, the Professor, play a game. George recorded all the shots he made (using a 2 for a two point shot and a 3 for a three point shot). At the end of the game, the Professor had a total of 20 points (coming from the two point and three point shots). George noticed in his list that every 3 was separated by at least one 2. How many different lists could George have written down?

Problem 12

Consider integers (x,y) that satisfy the equation

$$(x+y)^4 - 6xy(x^2+y^2) - 4x^2y^2 = 40.$$

How many such pairs of (x,y) exist?

Problem 13

A point is randomly chosen inside the square shown below.

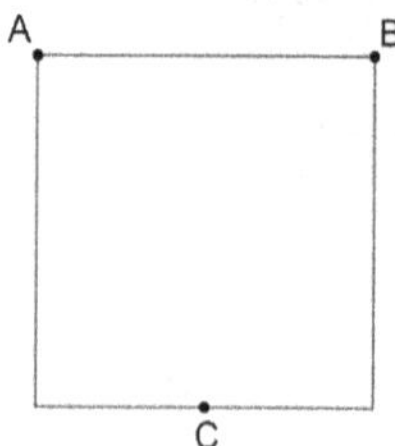

The probability that the chosen point is closest to point A can be written as $K\%$. What is K, rounded to the nearest integer if necessary?

Problem 14

$\triangle ABC$ is such that $AB = \sqrt{3}$ and $\angle ABC = 60^\circ$. A unit circle is drawn intersecting the triangle at two points D and E as shown in the diagram below.

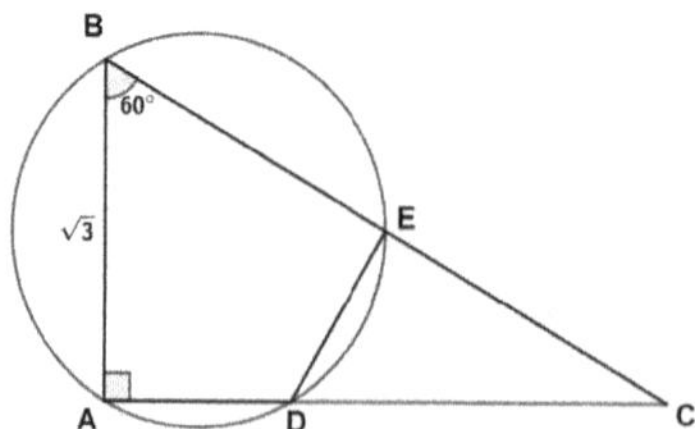

What is the area of quadrilateral $ABED$, rounded to the nearest tenth if necessary?

Problem 15

Consider the sequence $a_n = n \cdot 2^n + 1$. How many of the first 100 terms of this sequence $(a_0, a_1, \ldots, a_{99})$ have remainder 0 when divided by 3?

Problem 16

Let a, b, c, d be the roots of $2x^4 + 3x^3 - 6x^2 - 4x + 5 = 0$. Then

$$\frac{1}{a} + \frac{1}{b} + \frac{1}{c} + \frac{1}{d} = \frac{P}{Q}$$

for integers P, Q with $Q > 0$ and $\gcd(P, Q) = 1$. What is $P - Q$?

Problem 17

Let $ABCD$ be a unit square and let E, F be on sides $\overline{AD}, \overline{AB}$ respectively such that $\triangle AEF$ has perimeter 2 and area of $\triangle CEF$ is $\frac{17}{40}$. The area of $\triangle AEF$ can be written as $\frac{P}{Q}$ for positive integers P, Q with $\gcd(P, Q) = 1$. What is $P + Q$?

Problem 18

Let x be a real number such that $x^4 + 2x = 17$. Determine the value of $x^7 + 2x^4 - 17x^3$. Round your answer to the nearest integer if necessary.

Problem 19

Start with 5 slips of paper, 3 colored green and 2 colored blue. Randomly choose 3 of the slips of paper and throw away the other two. Cut the 3 chosen slips in half (resulting in 6 slips of paper). Again randomly choose 3 of these slips. The probability that these 3 slips are all green can be written as $\frac{P}{Q}$ for positive integers P, Q with $\gcd(P, Q) = 1$. What is $Q - P$?

Problem 20

A group of pirates went to hunt for treasure. They found a chest of gold coins. They tried to equally divide the coins, but 5 coins were left over. So they picked one pirate among themselves and threw him overboard. Then they tried to divide the coins again, but now 10 coins were left over! So again they picked another pirate among themselves and threw him overboard and tried to divide the coins again. This process continued until the coins could be evenly divided. If the chest held 100 coins, how many coins did each remaining pirate get when this process was completed?

1.8 ZIML May 2018 Junior Varsity

Below are the 20 Problems from the Junior Varsity ZIML Competition held in May 2018.
The answer key is available on p.180 in the Appendix.
Full solutions to these questions are available starting on p.138.

Problem 1

The equation $(x^2-4x+3)(x^2-1)=5$ has two real solutions. These two real solutions can be written in the form $A\pm\sqrt{B}$ where A and B are integers. What is $A+B$?

Problem 2

When asked about their favorite, Phil said he likes all even numbers, Roger said he likes all numbers made up of only even digits, Steve said he likes all the perfect squares, and Tony said he likes all multiples of 9. Consider the numbers $1,2,3,\ldots,100$. How many of them are liked by at least one of Phil, Roger, Steven, or Tony?

Problem 3

One edge of a regular octagon is extended to a line as shown in the diagram below.

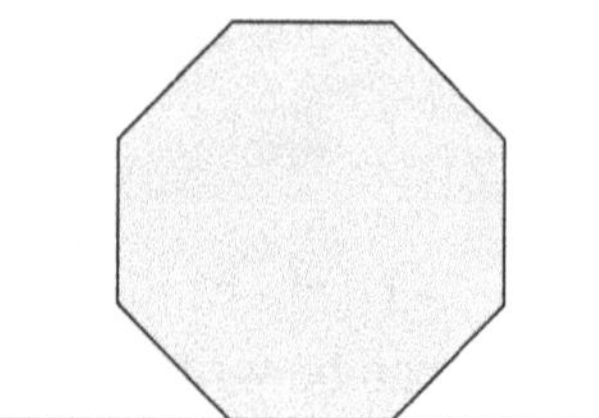

Consider lines that pass through two other vertices of the octagon and intersect the line above outside the octagon. What is the smallest angle formed by one of these intersections? Give your answer in degrees, rounded to the nearest hundredth if necessary.

Problem 4

Calculate the remainder when $1^3+2^3+3^3+\cdots+99^3$ is divided by 49.

Problem 5

Emily brings 12 tamales to a potluck with her friends Ania, Bridget, Carlos, and Duncan. She does not pay close attention at the party but definitely sees Ania eat one tamale and Duncan eat 4 tamales. At the end of the night each of the four friends shares how many tamales they eat (assume everyone eats an integer number of tamales). If the four friends do not necessarily eat all the tamales, how many different ways can they respond?

Problem 6

Find the sum of all solutions to $|11-|11-|11-2x|||=2$.

Problem 7

A and B are integers such that $\operatorname{lcm}(A,B)=6615$ and A has a total of 8 divisors. What is the greatest common factor of all such integers B?

Problem 8

Let $\triangle ABC$ be a 30-60-90 triangle with $\angle A=90^\circ$, $\angle B=60^\circ$, and $\angle C=30^\circ$, inscribed inside a circle. Let D be the midpoint of $\overline{AC}$ and extend $\overline{BD}$ to intersect the circle at E. If $AB=1$, the length DE can be written as $\frac{R\sqrt{S}}{T}$ for positive integers R,S,T with $\gcd(R,T)=1$ and S square-free. What is $R+S+T$?

Problem 9

Let r,s,t be the three roots of $x^3-4x^2+2x+4=0$. The expression

$$\left(\frac{r}{s}+\frac{r}{t}+1\right)+\left(\frac{s}{r}+\frac{s}{t}+1\right)+\left(\frac{t}{r}+\frac{t}{s}+1\right)$$

can be simplified and written as $\frac{P}{Q}$ for integers P,Q with $Q>0$ and $\gcd(P,Q)=1$. What is $P+Q$?

Problem 10

$ABCD$ is a parallelogram with area 180 (denoted $[ABCD] = 180$). Let E be a point on $\overline{BC}$ such that $BE = 2 \cdot CE$ and F be the intersection of $\overline{AE}$ with $\overline{BD}$. Calculate $[DEF]$ (the area of $\triangle DEF$). Round your answer to the nearest integer if necessary.

Problem 11

Find the remainder when 3^{2018} is divided by 31.

Problem 12

Derek has a deck of 10 cards, 5 of them red and 5 of them black. He deals 5 of them randomly to Alex, and Alex discards one card (of his choosing) so he ends up with 4 cards. Assuming Alex's goal is to have 4 cards of the same color, the probability Alex achieves his goal can be written as $\frac{P}{Q}$ for positive integers P, Q with $\gcd(P, Q) = 1$. What is $Q - P$?

Problem 13

Solve the equation $\sqrt{3x-5} - \sqrt{5x+1} = -2$. What is the sum of all the real solutions? Round your answer to the nearest tenth if necessary.

Problem 14

$\triangle ABC$ is a right triangle with $\angle B = 90^\circ$, $BC = 5$, and $AB = 12$. Let point D be on the extension of $\overline{BC}$ so that $\angle BAC = \angle CAD$ (so C is on line segment $\overline{BD}$). The length of AD can be expressed as the fraction $\frac{N}{M}$ where N, M are positive integers with $\gcd(N, M) = 1$. What is $N + M$?

Problem 15

Find the largest solution to $\lfloor 3x - 3 \rfloor = 2x + 1$. Round your answer to the nearest hundredth if necessary.

Problem 16

Mark has a bag containing 3 red golf balls (numbered $1, 2, 3$), 2 white golf balls (numbered $1, 2$), and 1 blue golf ball (numbered 1). He has randomly used one ball from the bag each of the last 6 times he's been golfing and always puts the ball back in the bag when he's done. Mark remembers he used 1 red, 2 white, and 3 blue golf balls (but not their numbers). How many different ways could he have chosen the golf balls?

Problem 17

Sarah and Elizabeth are on the school's soccer team. Assume when Sarah runs a mile it takes her between 6 and 8 minutes, and every time in-between is equally likely. Similarly assume when Elizabeth runs a mile it takes her between 7 and 10 minutes. They race by running a mile, with Sarah giving Elizabeth a 75 second head start. The probability Sarah wins the race can be expressed as $K\%$. What is K rounded to the nearest integer (if necessary)?

Problem 18

The quadratic equation $x^2+3kx+3k^2=27$ has real root(s) for x. For how many integer values of k are none of these roots negative?

Problem 19

What is the largest 5-digit number, with distinct digits, that leaves a remainder of 1 when divided by 9 and a remainder of 2 when divided by 11?

Problem 20

A cube and square right pyramid share a base $ABCD$. The pyramid and cube have the same volume with $K\%$ of the pyramid's volume contained inside the cube, where $K>0$. What is K, rounded to the nearest integer if necessary?

1.9 ZIML June 2018 Junior Varsity

Below are the 20 Problems from the Junior Varsity ZIML Competition held in June 2018.
The answer key is available on p.181 in the Appendix.
Full solutions to these questions are available starting on p.148.

Problem 1

For the equation $x^2 + mx + 3 = 0$, the difference between the two roots is 6 and the product of the two roots is n. What is $m^2 + n^2$? Round your answer to the nearest integer if necessary.

Problem 2

Start with equilateral triangle $\triangle ABC$. Sharing side $\overline{BC}$ construct a square. Then, sharing the side of this square containing point B, construct a regular pentagon.

Continue this pattern constructing a regular hexagon, etc., up to a regular decagon. Let $\overline{BE'}$ denote the side of this decagon that is not a side of the nonagon (9-sided shape). What is the measure of angle $\angle ABE'$, where $0^\circ \leq \angle ABE' < 180^\circ$? Round your answer to nearest whole degree.

Problem 3

A class of 4 boys and 6 girls stand in a circle and join hands. How many ways are there to arrange the 10 students so that each boy is separated by at least one girl? Note, we do not distinguish positions in the circle, just the arrangement of the students.

Problem 4

How many of the first 1000 squares, $1^2, 2^2, 3^2, \dots 1000^2$ have a remainder of 1 when divided by 12?

Problem 5

Tim's favorite authors are Asimov, Clarke, and Heinlein. He has 8 books by Asimov, 6 by Clarke, and 12 by Heinlein. He randomly arranges them in a line on his bookshelf, but makes sure to have the books grouped by author (so all the Asimov books are together, etc.).

The probability that the books are also arranged alphabetically by author can be written as $\frac{P}{Q}$ for positive integers P, Q with $\gcd(P,Q) = 1$. What is $P+Q$? (Here alphabetically by author means all the books by Asimov come before the books by Clarke, whose books all come before those by Heinlein.)

Problem 6

Let x be a real number satisfying $2x^4 + 5x^3 - 8x^2 + 5x + 2 = 0$. Calculate the value of $x + \frac{1}{x}$.

Problem 7

$\triangle ABC$ has sides $AB = 9$, $BC = 8$, and $AC = 7$. Points D, E, F are evenly spaced on $\overline{AB}$ and G, H are evenly spaced on $\overline{FC}$. (That is, $AD = DE = EF = FB$ and $FG = GH = HC$.) The area of $\triangle AGH$ can be expressed as $R\sqrt{S}$ for positive integers R and S where S contains no squares as factors. What is $R+S$?

Problem 8

Find the smallest 6-digit number that is divisible by 7, 9, 11, and 13.

Problem 9

A total of 4 line segments intersect as in the diagram below (the circle has center O):

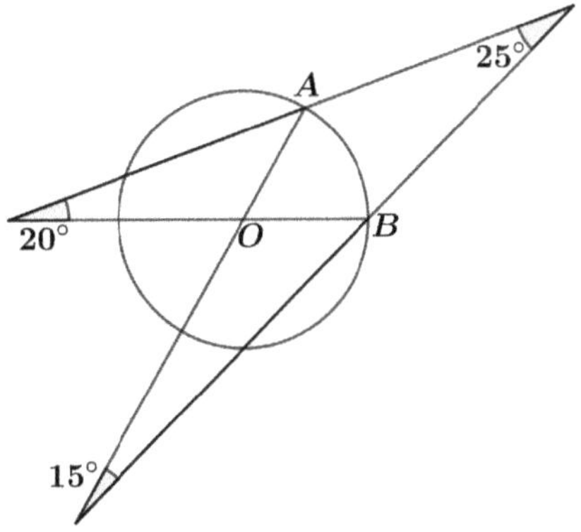

What is the measure of angle $\angle AOB$? Give your answer in degrees, rounded to the nearest integer if necessary.

Problem 10

Jaki was looking at her brother's homework assignment and noticed that she wrote

$$53+35=121$$

in one of her problems and it was marked correct by the teacher. Jaki then found out that her brother was learning about numbers in other bases. In what base were these numbers written so that the equation is true?

Problem 11

Norman went to the store to buy some soda for a party. He bought 8 bottles of soda in total, chosen from Sprite, Coke, Fanta, and Root Beer. If Norman did not buy the same number of bottles of Sprite and Coke, how many different ways could he have bought the soda? (Norman does not buy the soda in any particular order).

Problem 12

Consider solutions to the equation

$$x^2+2x-6=\sqrt{2x^2+4x+3}.$$

What is the (positive) difference between the largest real root and the smallest real root, rounded to the nearest integer if necessary.

Problem 13

Consider the expansion of $(x^3+2x-1)^3$, giving

$$x^9+ax^8+bx^7+cx^6+\cdots+hx+i.$$

What is the value of the expression

$$63a+127b+63c+31d+15e+7f+3g+h?$$

Problem 14

The three sides of a triangle are consecutive even integers. The angle bisector of the largest angle splits the opposite side into two parts, where the shorter part has length 8.4. What is the perimeter of the triangle?

Problem 15

Consider a list of all the factors of the number 60^{24}. The product of all the numbers in this list that are perfect squares or perfect cubes can be expressed as 60^K for an integer K. What is K?

Problem 16

Frankie and his classmate Chloe are learning about the Fahrenheit and Celsius temperature scales in chemistry class. Recall that the conversion from temperature in Fahrenheit (F) to temperature in Celsius (C) is given by

$$F = \frac{9}{5}C + 32.$$

Frankie picks a number randomly on a number line from 0 to 100 and Chloe does the same. The probability that Frankie's temperature, in Fahrenheit, is greater than Chloe's temperature, in Celsius, can be written as $K\%$. What is K, rounded to the nearest integer?

Problem 17

Consider pairs (x, y) of real numbers satisfying

$$|x| - y = 4 \text{ and } |x| \cdot y + x^2 = 0.$$

Over these pairs, what is the maximum value of $x \cdot y$? Round your answer to the nearest tenth if necessary.

Problem 18

Six congruent spheres with radius 1 are arranged so that the centers of the six spheres form a regular hexagon and each adjacent sphere is tangent. (Hence each of these spheres is tangent to two others.) A seventh sphere is tangent to all six spheres as well as the plane containing the hexagon described above. What is the radius of the seventh sphere? Round your answer to the nearest tenth if necessary.

Problem 19

Lost Larry starts at his house and goes exploring. He randomly walks 4 blocks, with each block being one block north or one block east. After these 4 blocks he tries to get back home. He knows he must walk south and west, but he doesn't remember how many of each. Thus, to try to get home he randomly walks another 4 blocks, with each block being one block south or one block west. How many of the $2^8 = 256$ paths in total result in Larry returning home?

Problem 20

Consider solutions to $x^2 - 5y = 4$ where x and y are integers. How many such solutions are there with $0 \leq x \leq 100$ and $0 \leq y \leq 100$?

2. ZIML Solutions

This part of the book contains the official solutions to the problems from the nine Junior Varsity ZIML Contests from the 2017-18 School Year.

Students are encouraged to discuss and share their own methods to the problems using the Discussion Forum on ziml.areteem.org.

2.1 ZIML October 2017 Junior Varsity

Below are the solutions from the Junior Varsity ZIML Competition held in October 2017.
The problems from the contest are available on p.17.

Problem 1 Solution

There are $5! = 120$ total outcomes for the boys. As we know the girls do not finish in a row, they must finish in the 6 spaces created by the boys (before all the boys, between the first and second boy, etc., to after all the boys). This gives $6 \cdot 5 \cdot 4 \cdot 3 = 360$ ways the girls can finish relative to the boys. In total we have $120 \cdot 360 = 43200$ outcomes.

Answer: 43200

Problem 2 Solution

The probability that Peter gets k tails is given by the formula

$$\frac{\binom{3}{k}}{2^3} \text{ for } k = 0, 1, 2, 3.$$

Similarly the probability that John gets k tails is given by the formula

$$\frac{\binom{5}{k}}{2^5} \text{ for } k = 0, 1, 2, 3, 4, 5.$$

For Peter and John to get the same number of tails, they can either get 0, 1, 2, or 3 tails, with probability

$$\frac{\binom{3}{0}\binom{5}{0}}{256} + \frac{\binom{3}{1}\binom{5}{1}}{256} + \frac{\binom{3}{2}\binom{5}{2}}{256} + \frac{\binom{3}{3}\binom{5}{3}}{256} = \frac{1+15+30+10}{256} = \frac{7}{32}.$$

Hence $P + Q = 7 + 32 = 39$.

Answer: 39

Problem 3 Solution

Let $AD = x$. Note that since $\overline{DE}$ is parallel to $\overline{BC}$ we have that $\triangle ADE$ is also isosceles so $AE = x$. Hence the perimeter of ADE is $2x + DE$ and the perimeter of $DECB$ is $2(12-x)+4+DE$. Hence $2x = 2(12-x)+4$ so $4x = 28$ and hence $x = AD = 7$.

Answer: 7

Problem 4 Solution

Factoring we have $x(x-1)(x+5)(x+6) = 72$. Regrouping we can write this as

$$(x)(x+5)(x-1)(x+6) = (x^2+5x)(x^2+5x-6) = 72$$

so if $y = x^2+5x$ we have $y(y-6) = 72$ or

$$y^2 - 6y - 72 = (y-12)(y+6) = 0.$$

If $y = 12$ we have $x^2+5x-12 = 0$ which has no integer solutions. If $y = -6$ we have

$$x^2+5x+6 = (x+3)(x+2) = 0$$

so $x = -3$ and $x = -2$ are integer solutions. Hence the sum of all integer solutions is $-3-2 = -5$.

Answer: -5

Problem 5 Solution

The largest possible number 98765 is not divisible by 9. If we try to change only the last digit, it doesn't work. Hence try the number $\overline{987ab}$. We need $a+b+7+8+9 = a+b+24$ to be divisible by 9. Thus $a+b \equiv 3 \pmod 9$ with a, b chosen from $0, 1, 2, \ldots, 6$ with b odd. As we want a to be as large as possible with b odd, we have $a = 2$, $b = 1$. Hence the largest number is 98721.

Answer: 98721

Problem 6 Solution

Since AE is the angle bisector, we have $AB : AC = BE : EC = 4 : 5$. Let $AB = 4x$ and $AC = 5x$. Then $4x + 5x + 8 + 10 = 72$ so solving for x we have $x = 6$. Therefore $AC = 5 \cdot 6 = 30$.

Answer: 30

Problem 7 Solution

We want to write

$$2000 = 2^4 \cdot 5^3 = (2^m \cdot 5^n)(2^p \cdot 5^q)(2^r \cdot 5^s),$$

the product of 3 factors. Thus we need $m+p+r=4, n+q+s=3$ with all numbers non-negative. Using stars and bars we have $\binom{4+3-1}{4} = 15$ ways to choose m, p, r and $\binom{3+3-1}{3} = 10$ ways to choose n, q, s. Hence there are $15 \cdot 10 = 150$ representations in total.

Answer: 150

Problem 8 Solution

Using the cross-multiplication formula with $a = b = 1$, we can assume there are positive integers c and d such that

$$x^2 - nx - 24 = (x - c)(x + d)$$

with $c > d$ and $cd = 24$. Then $n = c - d > 0$. Noting that

$$24 = 6 \cdot 4 = 8 \cdot 3 = 12 \cdot 2 = 24 \cdot 1,$$

we get $n = 6 - 4 = 2$, $n = 8 - 3 = 5$, $n = 12 - 2 = 10$, or $n = 24 - 1 = 23$, so the sum is $2 + 5 + 10 + 23 = 40$.

Answer: 40

Problem 9 Solution

All Pythagorean triples can be written in the form

$$(m^2 - n^2, 2mn, m^2 + n^2).$$

Note $2mn = 29$ is impossible. Further if $m^2 + n^2 = 29$ we must have $m, n = 5, 2$ which gives the triple $(20, 21, 29)$. Hence we need

$$m^2 - n^2 = (m - n)(m + n) = 29.$$

Since 29 is prime, we have $m - n = 1$ and $m + n = 29$, so $m = 15$ and $n = 14$. This gives the triple $(29, 420, 421)$ with sum

$$29 + 420 + 421 = 870.$$

Answer: 870

Problem 10 Solution

First we have $2017 \equiv 17 \pmod{20}$, so we need to calculate $17^{2018} \pmod{20}$. Finding a pattern we have $17^1 \equiv 17 \pmod{20}$, $17^2 \equiv 289 \equiv 9 \pmod{20}$, $17^3 \equiv 9 \cdot 17 \equiv 153 \equiv 13 \pmod{20}$, and $17^4 \equiv 13 \cdot 17 \equiv 221 \equiv 1 \pmod{20}$. Therefore, as 2016 is a multiple of 4, $17^{2018} \equiv 17^2 \equiv 9 \pmod{20}$.

Answer: 9

Problem 11 Solution

Note since the triangle has all 3 vertices on the square, the smallest possible side length is 4. Hence the area is equal to $4^2\sqrt{3}/4 = 4\sqrt{3} = \sqrt{48}$ so $K = 48$.

Answer: 48

Problem 12 Solution

Consider the following diagram

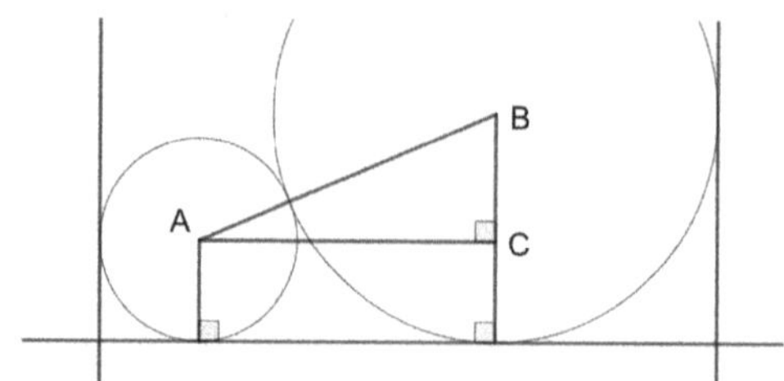

where A and B are the centers of C_1 and C_2. We have $AB = 4+9 = 13$ and $BC = 9-4 = 5$. Hence $\triangle ABC$ is a $5,12,13$ right triangle, so $AC = 12$. Therefore $|b-a|$, the horizontal distance between the two lines $x = a$ and $x = b$ is equal to $4+12+9 = 25$.

Answer: 25

Problem 13 Solution

Call the entire triangle ADE (with C on $\overline{AE}$). $\triangle ABC, \triangle ABE$ have the same height, so $[ABC] : [ABE] = 3 : 5 = 12 : 20$ (here $[ABC]$ is the area of $\triangle ABC$). Similarly, $[ABE] : [ADE] = 4 : 5 = 20 : 25$. Combining these two we have $[ABC] : [ADE] = 12 : 25 = 48 : 100$. Hence 48% of the diagram is shaded, so our answer is 48.

Answer: 48

Problem 14 Solution

As $0 \leq \{x\} < 1$ we must have $\lfloor x \rfloor > 4$. Trying $\lfloor x \rfloor = 5$, we have

$$5 \cdot \{x\} = 4 \Rightarrow \{x\} = \frac{4}{5} \Rightarrow x = 5 + \frac{4}{5} = \frac{29}{5}.$$

This is the smallest value, so $p+q = 29+5 = 34$.

Answer: 34

Problem 15 Solution

Using the quadratic formula,

$$x_{1,2} = \frac{m \pm \sqrt{m^2+32}}{2},$$

we see the positive difference of the roots is $\sqrt{m^2+32}$, so

$$\sqrt{m^2+32} = 2-m.$$

Hence

$$m^2+32 = m^2-4m+4,$$

so $-4m = 28$ and thus $m = -7$.
Double checking, the equation $x^2+7x-8=0$ has roots 8 and -1, with difference $8-(-1)=9$, which is exactly $2-(-7)$. As this is the only possible m, the final answer is -7.

Answer: -7

Problem 16 Solution

Note that the order of the rings on each finger matters! Hence the question is equivalent to the number of ways to order 8 rings, which is $8! = 40320$.

Answer: 40320

Problem 17 Solution

We have two cases: (i) all 3 numbers are odd, or (ii) one is odd and the other two even. Note 3 of the numbers on a 5-sided die are odd and 2 are even. Hence (i), we have $3\cdot 3\cdot 3 = 27$ outcomes. For (ii), any of the 3 rolls could be odd, so there are $3\cdot(3\cdot 2\cdot 2) = 36$ additional outcomes. This gives a total of $27+36 = 63$ outcomes with an odd sum.

Answer: 63

Problem 18 Solution

We have $6,250,000 = 50^4 = 2^4 \cdot 5^8$. Hence factors of $6,250,000$ that are perfect squares have the form $2^k \cdot 5^j$ where k is $0,2,4$ and $j = 0,2,4,6,8$. Hence there are $3 \cdot 5 = 15$ perfect square factors.

Answer: 15

Problem 19 Solution

The graph intersects the x-axis when $|x^2 - 2x - 3| = |x^2 + x - 2|$. This occurs either when $(x^2 - 2x - 3) = (x^2 + x - 2)$ or when $(x^2 - 2x - 3) = -(x^2 + x - 2)$. In the first case we have $-1 = 3x$ so $x = -1/3$. In the second case we want to solve $2x^2 - x - 5 = 0$. The discriminant is $(-1)^2 - 4 \cdot 2 \cdot (-5) > 0$ so this equation has 2 real solutions. Hence the graph crosses the x-axis $1 + 2 = 3$ times.

Answer: 3

Problem 20 Solution

By Vieta's formulas, $x_1 + x_2 = 10 - a$ and $x_1 x_2 = a$, so

$$x_1 x_2 + x_1 + x_2 = 10.$$

Completing the rectangle we have

$$(x_1 + 1)(x_2 + 1) = 11.$$

Since 11 is a prime, it can be factored as 1×11 or $(-1)(-11)$. So the two roots are 0 and 10, or -2 and -12. Hence a is either 0 or 24, but we want a positive integer, so the answer is 24.

Answer: 24

2.2 ZIML November 2017 Junior Varsity

Below are the solutions from the Junior Varsity ZIML Competition held in November 2017.
The problems from the contest are available on p.23.

Problem 1 Solution

Pairing up the factors, we see that n has a total of 4 factors. As $4 = 4 \times 1 = 2 \times 2$, n must be of the form p^3 or pq with $p > q$. The smallest of the form p^3 is $5^3 = 125$.

Hence we look at the form pq with $p > q$. For $q = 2$, we need $p > 50$ so the smallest is $p = 53$ so $pq = 106$. For $q = 3$, we need $p > 34$ so $p = 37$ and $pq = 111$. Proceeding we can fill in the following table:

q	2	3	5	7	11	13
smallest p	53	37	23	17	13	
p^2q	106	111	115	119	143	

Note for all $q \geq 13$ (since we are assuming $p > q$) the number will be larger than $13^2 = 169$. Hence we know that 106 is the smallest 3-digit number with 4 factors.

Answer: 106

Problem 2 Solution

Note the roots of $(2x^2+9x+4)^3 = 0$ are the roots of $2x^2+9x+4 = 0$ each repeated three times. Therefore we can cube the product of the roots of $2x^2+9x+4 = 0$. By Vieta's theorem, the product of the roots of $2x^2+9x+4 = 0$ is $\frac{4}{2} = 2$, so the product of the roots of $(2x^2+9x+4)^3 = 0$ is $2^3 = 8$.

Answer: 8

Problem 3 Solution

Substituting $y = \sqrt{x+2}$ we have

$$\frac{1}{y} + y = \frac{10}{3}.$$

Clearing denominators we have $3 + 3y^2 = 10y$ or

$$3y^2 - 10y + 3 = (3y-1)(y-3) = 0$$

so $y = 3$ or $y = 1/3$. If $\sqrt{x+2} = 3$ we have $x = 7$, which is not of the correct form. If $\sqrt{x+2} = 1/3$ we have $x = -17/9$ so $P - Q = -17 - 9 = -26$.

Answer: -26

Problem 4 Solution

Since $7102 \div 20 = 355.1$, there are 355 numbers between 1 and 7102 divisible by 20. Similarly there are 100 between 1 and 2017 divisible by 20 (as $2017 \div 20 = 100.85$). Hence there are $355 - 100 = 255$ numbers between 2017 and 7102 divisible by 20. This number includes numbers also divisible by 17.

Multiples of $\text{lcm}(20, 17) = 340$ are divisible by both 20 and 17. A method identical to above gives $20 - 5 = 15$ numbers divisible by 340 from 2017 to 7102.

Hence in total there are $255 - 15 = 240$ numbers from 2017 to 7102 divisible by 20 but not by 17.

Answer: 240

Problem 5 Solution

We will use the Principle of Inclusion-Exclusion (PIE). Let A represent the ways we can arrange the numbers so that 2 is next to 1, and B the ways we can arrange the numbers so that 2 is next to 3. Then $n(A) = n(B) = 2! \cdot 5! = 240$ and $n(A \cap B) = 2! \cdot 4! = 48$.

Thus there are

$$n(A)+n(B)-n(A\cap B)=240+240-48=432$$

total ways of doing this.

Answer: 432

Problem 6 Solution

Draw a height from side AB to side CD through P. Note this is combined height of the two triangles ABP and CDP. Hence the combined areas of the triangles equal half the area of the parallelogram, $40\div 2=20$. As $\triangle ABP$ has area 7, $\triangle CDP$ has area $20-7=13$.

Answer: 13

Problem 7 Solution

Let the number have digits a,b,c,d,e (e the ones, d the tens, etc.). For the number to be divisible by 11 we must have that

$$a+c+e-b-d\equiv 0 \pmod{11}.$$

We know a,b,c,d,e are chosen from $0,2,4,6,8$. As all of these choices are even (and $|a+c+e-b-d|\leq 16$) we must have

$$a+c+e=b+d.$$

We want the smallest number, so let $a=2$. It is impossible for $b=0$ ($2+c+e>0+d$ from the remaining numbers), so try $c=0$. Hence

$$2+0+e=b+d.$$

This is possible when $e=8, b=4, d=6$ (b and d could be switched). Since we want the smallest number, we have 24068 as our answer.

Answer: 24068

Problem 8 Solution

Using Heron's formula we can calculate the area of the triangle (the semiperimeter is 22):

$$\begin{aligned}\sqrt{22(22-20)(22-13)(22-11)} &= \sqrt{22\cdot 2\cdot 9\cdot 11}\\ &= \sqrt{22^2\cdot 9}\\ &= 66.\end{aligned}$$

The shortest altitude h will be to the longest side 20, so $\frac{20h}{2}=66$ and hence $h=\frac{132}{20}=\frac{66}{10}=6.6$. Therefore, rounded to the nearest integer, the shortest altitude is 7.

Answer: 7

Problem 9 Solution

Regular pentagons have angles of 108°, so $\angle TAP = 108^\circ$. Similar reasoning for equilateral triangles gives $\angle BAT = 60^\circ$. If $\triangle TAB$ is drawn inside the pentagon we get

$$\angle BAP = 108^\circ - 60^\circ = 48^\circ.$$

If $\triangle TAB$ is outside the pentagon we get

$$\angle BAP = 108^\circ + 60^\circ = 168^\circ.$$

As these are the only two possibilities, the largest possible is 168°.

Answer: 168

Problem 10 Solution

Substitute $x = 2-|y|$ into $x|y|-3x^3=0$, get $2|y|-y^2-3y^3=0$. We do case analysis for $y\geq 0$ and $y<0$:

If $y\geq 0$, $2y-y^2-3y^3=0$, and so

$$y(y+1)(2-3y)=0,$$

thus $y = 2/3$.

If $y < 0$, $-2y - y^2 - 3y^2 = 0$, and there is no real solution.

Therefore the answer is $y = 2/3 \approx 0.67$.

Answer: 0.67

Problem 11 Solution

Arrange the red cards (only 1 way). Then place 2 black cards in between the cards (only 1 way) using $2 \cdot 7 = 14$ black cards. Then the remaining 2 cards can arranged in the 9 spaces created by the red cards (now including the endpoints) using stars and bars. Hence our answer is

$$\binom{2+9-1}{2} = \binom{10}{2} = 45.$$

Answer: 45

Problem 12 Solution

The entire interval $(-5, 15)$ has length 20. If $x^2 > 16$ then $|x| > 4$. Hence the number must be in the interval $(-5, -4)$ (of length 1) or in the interval $(4, 15)$ (of length 11). Hence the probability is $\frac{1+11}{20} = \frac{12}{20} = 60\%$ so $N = 60$.

Answer: 60

Problem 13 Solution

$x_1 + x_2 = 6 - a$ and $x_1 x_2 = a$, so

$$(x_1 + 1)(x_2 + 1) = x_1 + x_2 + x_1 x_2 + 1 = 7.$$

As 7 is prime we get that the two roots x_1, x_2 are 0 and 6, or -8 and -2. If the roots are $0, 6$ then $a = 0$, so the roots must be -8

and -2, so $a = 16$.

Answer: 16

Problem 14 Solution

Let $AE = x, CE = 4y$, so $EB = 4x, ED = 9y$. By Power of a Point, $4x^2 = 36y^2$, so solving for x/y we get $x/y = 3$. Hence, $\frac{AB}{CD} = \frac{5x}{13y} = \frac{5}{13} \cdot 3 = \frac{15}{13}$. Thus $P + Q = 28$.

Answer: 28

Problem 15 Solution

The probability of heads and tails is each $\frac{1}{2}$.

If you get heads, rolling the die twice there are $6^2 = 36$ outcomes. Of these, 5, $(2,6), (3,5), (4,4), (5,3), (6,2)$, have a sum of 8, so the probability is $\frac{5}{36}$.

If you get tails, rolling the die three times there are $6^3 = 216$ outcomes. Here we want 3 positive numbers adding up to 8, so using stars and bars there are $\binom{8-1}{3-1} = \binom{7}{2} = 21$ outcomes (note here all the numbers are ≤ 6 as $1 + 1 + 6 = 8$). This gives a probability for this case of $\frac{21}{216} = \frac{7}{72}$.

Combining we have a probability of

$$\frac{1}{2} \cdot \frac{5}{36} + \frac{1}{2} \cdot \frac{7}{72} = \frac{5}{72} + \frac{7}{144} = \frac{17}{144}.$$

Therefore $Q - P = 144 - 17 = 127$.

Answer: 127

Problem 16 Solution

The numbers $1-9$ (9 total) each contribute 1 digit. The numbers $10-99$ (90 total) each contribute 2 digits. The numbers $100-999$ (900 total) each contribute 3 digits. Finally, the numbers $1000-2017$ (1018 in total) each contribute 4 digits. This gives a total of $1\cdot 9+2\cdot 90+3\cdot 900+4\cdot 1018=6961$ digits.

Answer: 6961

Problem 17 Solution

Note that

$$2^{38}-1=2^{12}\times(2^{26}-1)+(2^{12}-1)$$

so

$$\gcd(2^{38}-1,2^{26}-1)=\gcd(2^{26}-1,2^{12}-1).$$

Continuing in this same manner we have

$$\begin{aligned}\gcd(2^{26}-1,2^{12}-1) &= \gcd(2^{14}-1,2^{12}-1)\\ &= \gcd(2^{12}-1,2^{2}-1)\\ &= \gcd(4095,3)\\ &= 3.\end{aligned}$$

Hence $\gcd(2^{38}-1,2^{26}-1)=3$.

In fact it is true for $m,n\geq 1$ that

$$\gcd(2^m-1,2^n-1)=2^{\gcd(m,n)}-1.$$

Answer: 3

Problem 18 Solution

First note $\sqrt{2x}>\sqrt{2x-10}$ so $K\geq 0$. Isolating $\sqrt{2x}$ we have

$$\begin{aligned}&\sqrt{2x}=\sqrt{2x-10}+K\\ \Rightarrow\ & 2x=2x-10+2K\sqrt{2x-10}+K^2\\ \Rightarrow\ & 2K\sqrt{2x-10}=10-K^2.\end{aligned}$$

Since $K \geq 0$ it is clear from this that $10 - K^2 \geq 0$ so $0 \leq K \leq \sqrt{10}$. This gives $K = 3$ as the largest integer. (To double check, note that $x = \dfrac{361}{72}$ is a solution when $K = 3$.)

Answer: 3

Problem 19 Solution

First note using Fermat's Little Theorem, $3^{10} \equiv 1 \pmod{11}$. Since $3^3 \equiv 27 \equiv 7 \pmod{10}$ we have

$$3^{3^3} \equiv 3^{27} \equiv 3^7 \equiv (3^3)^2 \cdot 3 \equiv 27^2 \cdot 3 \equiv 5^2 \cdot 3 \equiv 75 \equiv 9 \pmod{11},$$

so the remainder is 9.

Answer: 9

Problem 20 Solution

As $ABCD$ is a trapezoid, we know that

$$\angle A + \angle B = \angle C + \angle D = 180°.$$

Since $ABCD$ is cyclic, we know

$$\angle A + \angle C = \angle B + \angle D = 180°.$$

Subtracting we get that $\angle B = \angle C$ (so also $\angle A = \angle D$) so $ABCD$ is an isosceles trapezoid. Hence

$$BC = CD = DA = 4.$$

Let O be the center of the circle. Drawing OA, OB, OC, OD (all with length 4) note that $\triangle OAD \cong \triangle OBC \cong OCD$ and each are equilateral triangles. Thus O is on $\overline{AB}$ so $\overline{AB}$ is a diameter and $AB = 8$.

Drawing heights from C and D we divide the trapezoid into a rectangle and two right triangles with hypotenuse 4 and leg 2.

Hence these are $30-60-90$ triangles and the height of the trapezoid is $2\sqrt{3}$. This gives an area of

$$\frac{4+8}{2} \times 2\sqrt{3} = 12\sqrt{3} \approx 12 \times 1.73 = 20.76$$

so rounded to the nearest integer the area is 21.

Answer: 21

2.3 ZIML December 2017 Junior Varsity

Below are the solutions from the Junior Varsity ZIML Competition held in December 2017.
The problems from the contest are available on p.27.

Problem 1 Solution

First note $\overline{CD} \| \overline{AB}$ as the arcs $\widehat{AC}$ and $\widehat{DB}$ have the same measure. Therefore $\triangle ACD$ and $\triangle OCD$ have the same area, where O is the center of the circle. Hence the area of the figure we want is the same as the area of the sector with arc $\widehat{CD}$. Since $\widehat{CD}$ is $\frac{2}{3}$ of arc $\widehat{AB}$ it is a one-third of a circle with radius 6. Therefore the area we want is

$$\frac{1}{3} \cdot \pi \cdot 6^2 = 12\pi,$$

so $K = 12$.

Answer: 12

Problem 2 Solution

We must have $4 \mid \overline{7b}$, so $b = 2$ or 6. The sum of the digits is a multiple of 9, so

$$a+3+5+7+b \equiv a+b+6 \equiv 0 \pmod 9.$$

Since a, b are digits from $1-9$, if $b = 2$ then $a = 1$ while if $b = 6$ then $a = 6$. This gives the two numbers 13572 and 63576 with sum $13572 + 63576 = 77148$.

Answer: 77148

Problem 3 Solution

There are 3 choices for each of his four sons (total of $3^4 = 81$ possibilities). For the daughters, there are $4 \cdot 3 \cdot 2 = 24$ possibilities since they all must be at a different school. The final answer is $81 \cdot 24 = 1944$.

Answer: 1944

Problem 4 Solution

Note the equation can be rewritten $3y - 3x = xy$. Completing the rectangle, $(x-3)(y+3) = -9$. Then

$$\begin{aligned} -9 &= (1)(-9) &= (-1)(9) &= (3)(-3) \\ &= (-3)(3) &= (9)(-1) &= (-1)(9) \end{aligned}$$

leads to 6 solutions to $3y - 3x = xy$. ($(x-3, y+3) = (1, -9)$, $(-1, 9)$, etc.) However, one of these gives $(x, y) = (0, 0)$, which doesn't work as $x, y \neq 0$. The other 5 give the solutions

$$(-6, -2), (2, 6), (4, -12), (6, -6), (12, -4)$$

for (x, y).

Answer: 5

Problem 5 Solution

We have the equation

$$pqr = 19(p+q+r).$$

Since p, q, and r are all prime numbers, one of them must be 19. Without loss of generality, assume $p = 19$, then

$$19qr = 19(19+q+r),$$

so

$$qr = 19+q+r,$$

and

$$qr - q - r = 19.$$

Completing the rectangle,

$$qr - q - r + 1 = 20,$$

so

$$(q-1)(r-1)=20.$$

Since switching q and r does not give an additional solution, we assume $q \geq r$. There are 3 cases:
(1) $q-1=20$, $r-1=1$. So $q=21$, $r=2$, but q is not an prime in this case.
(2) $q-1=10$, $r-1=2$.
So we get $q=11$ and $r=3$, which is a solution.
(3) $q-1=5$, $r-1=4$, so $q=6$, but it is not a prime.

Therefore the only triple of such prime number is $(19,11,3)$, so

$$q^2+r^2+h^2=361+121+9=491.$$

Answer: 491

Problem 6 Solution

Call this number N. Note that $N+5$ is divisible by $3,5,7,9$, so the smallest $N+5$ will be $\text{lcm}(3,5,7,9)=315$. Hence $N=310$.

Answer: 310

Problem 7 Solution

Make the substitution $y=\sqrt{\frac{x-2}{x+2}}$ to get $y+\frac{3}{y}=4$. Solving for y gives $y=1,3$. Hence $\frac{x-2}{x+2}=1$ or 9. 1 is impossible, so $x-2=9(x+2)$ and hence $x=\frac{-5}{2}$. Therefore $P\cdot Q=-10$.

Answer: -10

Problem 8 Solution

With no order there are $\binom{12}{3}$ ways to choose 3 balls. Consider the three cases of all red, all green, all yellow, so there are

$$\binom{5}{3}+\binom{4}{3}+\binom{3}{3}$$

ways to get all 3 of the same color. Hence the probability is

$$\frac{\binom{5}{3}+\binom{4}{3}+\binom{3}{3}}{\binom{12}{3}}=\frac{15}{220}=\frac{3}{44}$$

so $Q-P=44-3=41$.

Answer: 41

Problem 9 Solution

Consider the diagram below with the circle drawn.

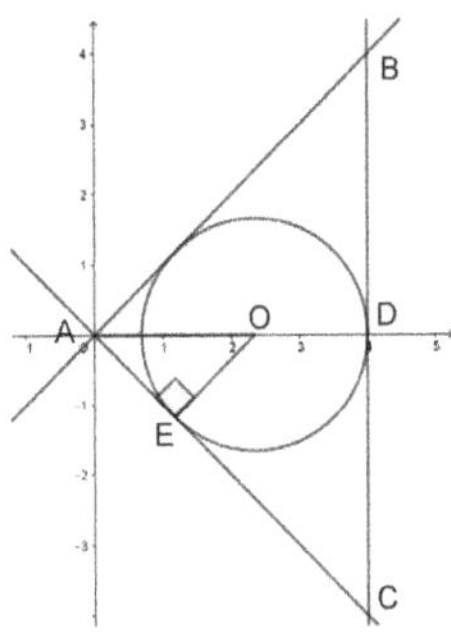

$\triangle ABC$ is a right triangle, and $AD=4$. Further, $\overline{OA}$ bisects $\angle CAB=90^\circ$, so $\triangle AOE$ is a $45-45-90$ triangle. Using r for the radius, we therefore have

$$\begin{aligned}4=AD&=AO+OD=r\sqrt{2}+r\\ \Rightarrow\quad r&=\frac{4}{\sqrt{2}+1}=4\sqrt{2}-4=\sqrt{32}-4.\end{aligned}$$

Hence $R+S=32-4=28$.

Answer: 28

Problem 10 Solution

Since the dart lands randomly inside the circles, the probability of winning in a single try is $\frac{\pi \cdot 2^2}{\pi \cdot 4^2} = \frac{1}{4}$. Thus the probability you win in total is

$$\frac{1}{2} \cdot \frac{1}{4} + \frac{1}{2} \cdot \left[1 - \left(\frac{3}{4}\right)^2\right],$$

as to win when getting tails you only need to succeed once. This gives a final answer of

$$\frac{1}{8} + \frac{7}{32} = \frac{11}{32} \approx 0.3437 \approx 34\%$$

so $W = 34$.

Answer: 34

Problem 11 Solution

As $\{x\} < 1$, $4\{x\} < 4$, we must have $\lfloor x \rfloor \geq 5$. Trying $\lfloor x \rfloor = 5$, we have

$$5 + 4\{x\} = 8 \Rightarrow \{x\} = 0.75.$$

Hence $x = 5 + 0.75 = 5.75$ is the smallest solution.

Answer: 5.75

Problem 12 Solution

Think of the cookies coming in pairs, so you want to buy 3 pairs of cookies. Using stars and bars, there are $\binom{3+8-1}{3} = 120$ ways of buying the cookies.

Answer: 120

Problem 13 Solution

If we let $z = (x+1)^2$ we have

$$4z^2 + 7xz - 2x^2 = (4z - x)(z + 2x) = 0.$$

Hence either $4(x+1)^2-x=0$ or $(x+1)^2+2x=0$. In the first case we have $4x^2+7x+4=0$ which has no real roots (discriminant is -17). In the second case we have $x^2+4x+1=0$ which has roots $-2\pm\sqrt{3}$ using the quadratic formula. Hence $A-B=-2-3=-5$.

Answer: -5

Problem 14 Solution

The first few terms of a_n are $1,1,3,7,17,41,99,239,\ldots$, which gives $1,1,3,3,1,1,3,3,\ldots$ modulo 4. We see that the sequence $a_n \pmod 4$ has period 4. Since $2018\equiv 2 \pmod 4$, we have $a_{2018}\equiv a_2\equiv 1 \pmod 4$. Hence the remainder is 1.

Answer: 1

Problem 15 Solution

Use $x^3=-9x+27$ to reduce the exponents:

$$\begin{aligned} x^7+486x^2 &= x(-9x+27)^2+486x^2 \\ &= 81x^3-486x^2+729x+486x^2 \\ &= 81(x^3+9x) \\ &= 81\times 27 \\ &= 2187. \end{aligned}$$

Answer: 2187

Problem 16 Solution

Consider the diagram below, with the height of the trapezoid drawn from O to $\overline{CD}$.

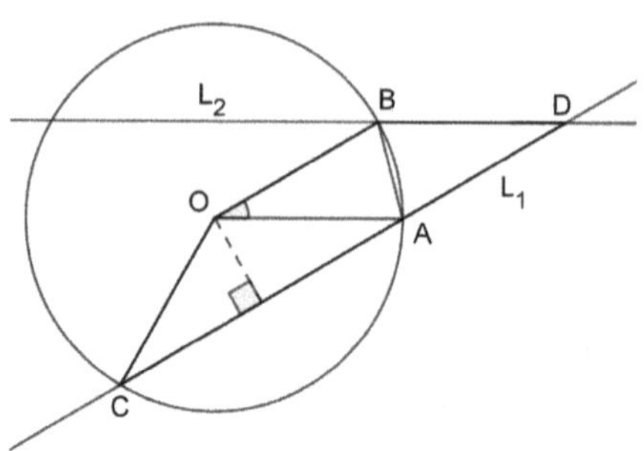

Since $AOBD$ is a parallelogram we have

$$OB = AD = OA = DB = 4.$$

Further,

$$\angle AOB = \angle OAC = 30^\circ,$$

so as $\triangle OAC$ is isosceles, $\angle OCA = 30^\circ$ and hence the drawn height divides $\triangle OAC$ into two 30-60-90 triangles. Thus the height is $4 \div 2 = 2$ and

$$AC = 2 \cdot 2\sqrt{3} = 4\sqrt{3}.$$

This gives an area for trapezoid $BOCD$ of

$$\frac{4 + 4\sqrt{3} + 4}{2} \cdot 2 = 8 + 4\sqrt{3} \approx 8 + 6.92 = 14.92$$

using $\sqrt{3} \approx 1.73$. Rounded to the nearest integer, the answer is 15.

Answer: 15

Problem 17 Solution

Examining one base and one side we have the diagram below:

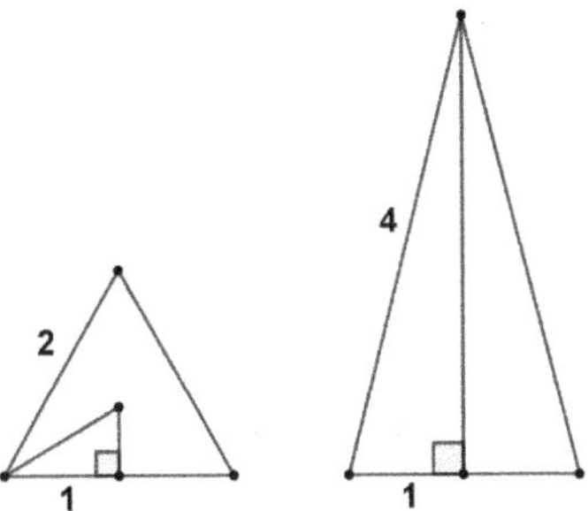

The left triangle is the base, which is an equilateral triangle. Hence the right triangle drawn, which gives the incenter of the base, is a 30-60-90 triangle with sides $\sqrt{3}/3, 1, 2\sqrt{3}/3$. We use the diagram on the right to find the slant height of the pyramid, which is $\sqrt{15}$. Thus if h is the height of the pyramid (recall the apex is directly above the incenter of the base) we have

$$\left(\frac{\sqrt{3}}{3}\right)^2 + h^2 = \sqrt{15}^2 \Rightarrow h^2 = \frac{44}{3} \Rightarrow h = \frac{2\sqrt{33}}{3}.$$

This gives a final volume of the pyramid of

$$\frac{1}{3} \cdot \frac{2^2\sqrt{3}}{4} \cdot \frac{2\sqrt{33}}{3} = \frac{2\sqrt{11}}{3} = \frac{\sqrt{44}}{3}.$$

Thus $K = 44$.

Answer: 44

Problem 18 Solution

For there to be real roots, the discriminant must be ≥ 0, hence

$$\begin{aligned}(2m)^2 - 4(12-m) \geq 0 \quad &\Rightarrow \quad m^2 + m - 12 \geq 0 \\ &\Rightarrow \quad m \geq 3 \text{ or } m \leq -4.\end{aligned}$$

Let r,s be the the two roots. We want $r,s \geq 1$, which means $r-1 \geq 0$ and $s-1 \geq 0$. Using Vieta's Theorem,

$$\begin{aligned}(r-1)+(s-1) \geq 0 \ &\Rightarrow \ r+s \geq 2 \\ &\Rightarrow \ -2m \geq 2 \\ &\Rightarrow \ m \leq -1.\end{aligned}$$

Similarly

$$\begin{aligned}(r-1)(s-1) \geq 0 \ &\Rightarrow \ rs-(r+s)+1 \geq 0 \\ &\Rightarrow \ 12-m-(-2m)+1 \geq 0 \\ &\Rightarrow \ m \geq -13.\end{aligned}$$

Combining all the restrictions we see $-13 \leq m \leq -4$ so there are 10 such integral m.

Answer: 10

Problem 19 Solution

Rotate $\triangle PDC$ 90° about C (so D is rotated to B). Suppose the point P is rotated to P'. By the rotation, $BP' = DP = 1$, and also

$$P'Q = BP' + BQ = PD + BQ = 1 + 2 = 3 = PQ.$$

Since $PC = P'C$ and $QC = QC$, we obtain $\triangle PQC \cong \triangle P'QC$ by SSS congruency. Hence $\angle P'CQ = \angle PCQ$. As $PCP' = 90^\circ$,

$$\angle PCQ = 90^\circ \div 2 = 45^\circ.$$

Answer: 45

Problem 20 Solution

Let A,B,C be events that each of the three friends get 0 books respectively, so we want the opposite of $A \cup B \cup C$. We have

$$n(A) = n(B) = n(C) = 2^5,$$

$$n(A \cap B) = n(A \cap C) = n(B \cap C) = 1^5,$$

and

$$n(A \cap B \cap C) = 0,$$

so using PIE (Principle of Inclusion-Exclusion),

$$n(A \cup B \cup C) = 3 \cdot 2^5 - 3.$$

To get our final answer we subtract this from 3^5, which is the total number of outcomes with no restrictions. Therefore our answer is

$$3^5 - 3 \cdot 2^5 + 3 = 150.$$

Alternatively the problem can be done directly by considering the case of 1 friend getting 3 books (and the others each 1 book) and the case of 2 friends getting 2 books (and the other getting 1 book) separately.

Answer: 150

2.4 ZIML January 2018 Junior Varsity

Below are the solutions from the Junior Varsity ZIML Competition held in January 2018.
The problems from the contest are available on p.33.

Problem 1 Solution

Since $9 \times 11 = 99$, this is equivalent to our number being divisible by 9 and having remainder 9 when divided by 11. To be divisible by 9 we need

$$8+3+2+x+y+1+5+9+9 \equiv x+y+1 \equiv 0 \pmod{9},$$

so $x+y \equiv 8 \pmod{9}$. To leave remainder 9 when divided by 11 we also need

$$9-9+5-1+y-x+2-3+8 \equiv y-x \equiv 9 \pmod{11}.$$

Since $x+y \equiv 8 \pmod{9}$ we must have $x+y=8$ or $x+y=17$. As $y-x \equiv 9 \pmod{11}$ we must have $y-x=9$ or $y-x=-2$. Note $y-x=9$ implies $y=9, x=0$ which is impossible. Hence $y-x=-2$ so $x-y=2$. This is only possible when $x+y=8$, which gives $x=5$ and $y=3$. Therefore $10x+y=53$.

Answer: 53

Problem 2 Solution

The hole has a radius of 10 and a depth of 6, giving the side view shown below:

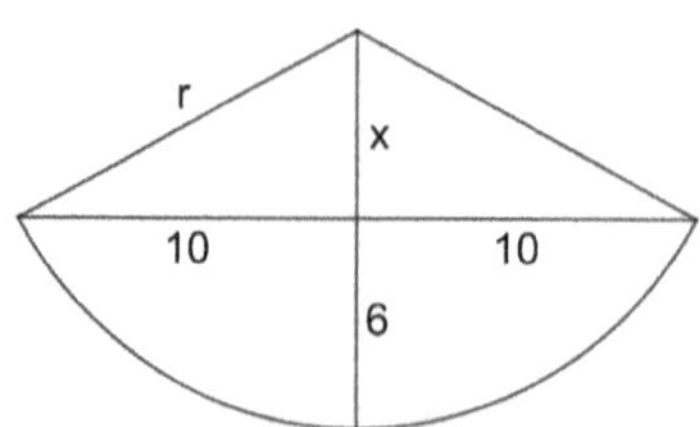

Note $r = x+6$ so $(x+6)^2 = x^2 + 10^2$ using the Pythagorean theorem. This gives $12x+36 = 100$ so $12x = 64$ and therefore $x = \frac{64}{12} = \frac{16}{3}$. Thus $r = \frac{16}{3} + 6 = \frac{34}{3}$ and $P+Q = 34+3 = 37$.

Answer: 37

Problem 3 Solution

Set up equation $10a+b = a+b^2$, so $9a = b(b-1)$. Thus either 9 divides b or 9 divides $(b-1)$, which means $b = 9, 0$, or 1. Only $b = 9$ works, which gives $a = 8$. Hence our number is 89.

Answer: 89

Problem 4 Solution

Let x be the distance from E to $\overline{CD}$ so $x+5$ is the distance from E to $\overline{BC}$. Since the diagonals intersect at the center of the rectangle, we see the rectangle has dimensions $2x$ inches by $2(x+5)$ inches. Thus we have

$$2x \cdot 2(x+5) = 56 \Rightarrow x^2 + 5x - 14 = 0.$$

Solving we have $x = -7$ and $x = 2$ so the rectangle must be 4 inches by 14 inches. Hence the perimeter is $2(4+14) = 36$ inches.

Answer: 36

Problem 5 Solution

Note the only way we will get terms with x^4 are $x^4 \times 4 \times (-3)$ or $3 \times x^2 \times x^2$. Hence, the coefficient of $x^4 = -12+3 = -9$.

Answer: -9

Problem 6 Solution

Since 11 is prime, we can use Fermat's Little Theorem to get $9^{10} \equiv 1 \pmod{11}$. Therefore,

$$9^{2018} \equiv 9^8 \equiv (-2)^8 \equiv 256 \equiv 3 \pmod{11}.$$

Answer: 3

Problem 7 Solution

First divide the couples into three groups: (i) the Smiths/Lees, (ii) the Johnsons, and (iii) the Garcias. There are 3! ways to arrange these three groups. For the Smiths/Lees there are 2! ways to arrange the two couples. Lastly, for each of the 4 couples there are 2! ways to arrange the two people in the couple. Multiplying all these together gives our final answer of

$$3! \cdot 2! \cdot (2!)^4 = 3 \cdot 2^6 = 192.$$

Answer: 192

Problem 8 Solution

Note that $\frac{1}{ab} + \frac{1}{ac} + \frac{1}{bc} = \frac{a+b+c}{abc}$. Using Vieta's formulas, we have $a+b+c = -\frac{-10}{1/5} = 50$ and $abc = -\frac{25}{1/5} = -125$. Hence $\frac{1}{ab} + \frac{1}{ac} + \frac{1}{bc} = \frac{50}{-125} = -\frac{2}{5} = -0.4$.

Answer: -0.4

Problem 9 Solution

We want to find $[BEF]/[ABCD]$, where $[BEF]$ denotes the area of $\triangle BEF$, and so on.

Since E is a midpoint,

$$[BED] = [ABCD]/4.$$

Since $ABCD$ is a parallelogram and $BE = AD/2$, $\triangle DAF \sim \triangle BEF$ with ratio of side lengths $2:1$. Using this information, we have $[DEF] = 2[BEF]$. Therefore,

$$[BEF] = [ABCD]/12.$$

Therefore $[BEF]$ is $\frac{1}{12} \approx 8.3\%$ of the entire parallelogram.

Answer: 8.3

Problem 10 Solution

The set has size 8, so in total there are $2^8 = 256$ subsets. There are 3 odd numbers, so $2^3 = 8$ subsets consist of only odd numbers. The other 5 numbers are even, hence there are $2^5 = 32$ subsets of only even numbers. Note, however, we are counting the empty subset in both of these cases. Therefore there are

$$256 - 8 - 32 + 1 = 217$$

subsets that contain at least one odd and at least one even number.

Answer: 217

Problem 11 Solution

We want x so that $x^2 \equiv 1 \pmod{12}$. Checking for

$$x = 0, 1, 2, \ldots, 11$$

we see that if x is equivalent to 1, 5, 7, or 11 modulo 12, then x^2 has remainder 1. As $96 = 8 \times 12$, there are $4 \times 8 = 32$ squares from 1^2 to 96^2 with remainder 1 when divided by 12. Among 97^2 to 100^2 there is one more 98^2, for a total of 33.

Answer: 33

Problem 12 Solution

The circle has radius 4 so we know $OA = OB = 4$ and $OC = 2$. Therefore $\triangle OCD$ is a 30-60-90 triangle so $\overline{OB} \perp \overline{AD}$ and $AC = 2\sqrt{3}$. Hence $\overline{AD}$ is perpendicular to a radius, so $CD = AC = 2\sqrt{3}$ and hence $AD = 4\sqrt{3} = \sqrt{48}$ and $L = 48$.

Answer: 48

Problem 13 Solution

Factoring we have

$$(x^2-1)(x^2+10x+24) = (x-1)(x+1)(x+4)(x+6).$$

Regrouping the Left Hand Side as $(x-1)(x+6)(x+1)(x+4)$, the equation becomes

$$(x^2+5x-6)(x^2+5x+4) = 24.$$

Making the change of variables $y = x^2+5x-6$, we have

$$y(y+10) = 24$$

, or

$$y^2+10y-24 = 0,$$

which is

$$(y+12)(y-2) = 0$$

so $y = 2$ or $y = -12$. Hence we have either $x^2+5x-6 = 2$ which has no integer solutions, or $x^2+5x-6 = -12$ which gives $x = -3$ or -2. Hence the product of all integer solutions is $(-3)(-2) = 6$.

Answer: 6

Problem 14 Solution

There are $2^8 = 256$ total outcomes. Of these, $\binom{8}{4} = 70$ have an equal number of heads and tails. Hence there are $256 - 70 = 186$ outcomes with an unequal amount of heads and tails. Half of these, $186 \div 2 = 93$ have more heads. Thus the probability is $\frac{93}{256}$ so $P + Q = 349$.

Answer: 349

Problem 15 Solution

Since G is the centroid of $\triangle ABC$, all the small triangles have the same area: $78 \div 6 = 13$. The shaded area consists of 2 of these triangles, so has area $K = 2 \times 13 = 26$.

Answer: 26

Problem 16 Solution

Note $-(x^2 - 2x - 3) + (x^2 + x - 2) = 3x + 1$ so we want x such that $x^2 - 2x - 3 \leq 0$ and $x^2 + x - 2 \leq 0$. We have

$$x^2 - 2x - 3 = (x+1)(x-3) \text{ and } x^2 + x - 2 = (x+2)(x-1),$$

so for both quadratics to be negative we need $-1 \leq x \leq 3$ and $-2 \leq x \leq 1$ or $-1 \leq x \leq 1$. Hence $L + H = -1 + 1 = 0$.

Answer: 0

Problem 17 Solution

First there are $\binom{9}{4}$ ways to choose which 4 friends sit with Carrie. The 4 friends then need to be arranged around the table with Carrie, which can be done in $4!$ ways. Similarly, the other 5 friends can be seating in $5! \div 5 = 4!$ ways around the second circular table. This gives a final answer of

$$\binom{9}{4} \cdot 4! \cdot 4! = \frac{9!}{5} = 72576.$$

Answer: 72576

Problem 18 Solution

Let $u = x + y$, $v = xy$, then

$$u^2 - 3v = 13,$$

and

$$u - v = -5.$$

Solve to get $(u, v) = (7, 12)$ or $(u, v) = (-4, 1)$. For each pair of (u, v), set up quadratic equation in t:

$$t^2 - ut + v = 0$$

where x, y are the two roots. Thus,

$$t^2 - 7t + 12 = 0,$$

or

$$t^2 + 4t + 1 = 0.$$

Therefore using Vieta's formulas,

$$x + y = -(-7) = 7 \text{ or } x + y = -4.$$

Hence, 7 is the largest such $x + y$.

Note: One can verify that $(3, 4)$, $(4, 3)$, $(-2 + \sqrt{3}, -2 - \sqrt{3})$, and $(-2 - \sqrt{3}, -2 + \sqrt{3})$ are all the possible pairs for (x, y).

Answer: 7

Problem 19 Solution

There is a $\frac{1}{6}$ chance of rolling a $1,2,3,4,5,6$. If you roll a one or a two, the probability the real number is less than 2 is 1. For $3,4,5,6$, the probability is respectively $\frac{2}{3},\frac{2}{4},\frac{2}{5},\frac{2}{6}$. Hence using the law of total probability we have

$$\frac{1}{6}\left(1+1+\frac{2}{3}+\frac{2}{4}+\frac{2}{5}+\frac{2}{6}\right)=\frac{1}{6}\cdot\frac{39}{10}=\frac{13}{20},$$

so $Q-P=7$.

Answer: 7

Problem 20 Solution

We have that $140=2^2\cdot 5\cdot 7$. Therefore x is either 2, 5, or 7.

First assume $x=7$, so we have $(y+7)(y+z)=20$. As $y,z\geq 2$, we have $(y+7)(y+z)\geq 9\cdot 4=36$ so this is impossible.

Next assume $x=5$, so we have $(y+5)(y+z)=28$. As $y,z\geq 2$, the first term must be at least 7 and the second term must be at least 4, leaving only $28=7\cdot 4$ which implies $y=2$ and $z=2$.

Lastly assume $x=2$, so we have $(y+2)(y+z)=70$. As $y,z\geq 2$, we look at the factor pairs: $(14,5)$, $(10,7)$, $(7,10)$, and $(5,14)$. Of these, only $(7,10)$ and $(5,14)$ work, leading to pairs $(5,5)$ and $(3,11)$.

Hence there are three triples in total: $(5,2,2)$, $(2,5,5)$, and $(2,3,11)$. Only $(2,3,11)$ has x,y,z distinct, giving an answer of $x+y+z=16$.

Answer: 16

2.5 ZIML February 2018 Junior Varsity

Below are the solutions from the Junior Varsity ZIML Competition held in February 2018.
The problems from the contest are available on p.39.

Problem 1 Solution

There are $2^4 = 16$ possibilities for boy/girl for each of the 4 children, but we remove the 2 possibilities of all girls or all boys as the family has at least one of each. Of these, $\binom{4}{1} + \binom{4}{2} = 10$ have at most two girls. This gives a probability of

$$\frac{10}{14} = \frac{5}{7} \approx 0.7143$$

or approximately 71%. Hence the answer K is 71.

Answer: 71

Problem 2 Solution

We in fact have that $\triangle ABO, \triangle BCO$ are both equilateral with side length 6. Hence the area of $ABCO$ is twice the area of equilateral triangle ABO:

$$2 \cdot \frac{6^2\sqrt{3}}{4} = 18\sqrt{3}.$$

Thus $P + Q = 18 + 3 = 21$.

Answer: 21

Problem 3 Solution

Consider the cases of $x \geq 0$ or $x < 0$.

If $x \geq 0$ we have $x + y = 2$ and $x + y^2 = 4$. Subtracting we get $y^2 - y = 2$ or $y^2 - y - 2 = (y-2)(y+1) = 0$ so $y = 2$ or $y = -1$. If $y = 2, x = 0$. If $y = -1, x = 3$.

If $x < 0$ we have $-x+y=2$ and $x+y^2=4$. Adding we get $y^2+y=6$ or $y^2+y-6=(y+3)(y-2)=0$ so $y=2$ or $y=-3$. We already had $y=2$, and if $y=-3$ we get $x=-5$. They are all the solutions.

The largest value of x^2+y^2 occurs for the pair $(-5,-3)$ giving $x^2+y^2=(-5)^2+(-3)^2=34$.

Answer: 34

Problem 4 Solution

We want to calculate $2016^{2017}+2020^{2017} \pmod{2018}$. Note we have $2016 \equiv -2 \pmod{2018}$ and $2020 \equiv 2 \pmod{2018}$. Hence

$$2016^{2017}+2020^{2017} \equiv (-2)^{2017}+2^{2017} \equiv 0 \pmod{2018}.$$

Thus the remainder is 0.

Answer: 0

Problem 5 Solution

If r is a root of $x^3-4x+3=0$ we have $r^3=4r-3$. Hence

$$a^3+b^3+c^3=4a-3+4b-3+4c-3=4(a+b+c)-9.$$

As $a+b+c=0$ using Vieta's theorem, we have

$$a^3+b^3+c^3=-9.$$

Answer: -9

Problem 6 Solution

Note that

$$\begin{aligned} \overline{abc}+\overline{bcd} &= 100a+110b+11c+d \\ &= a+d+11(9a+10b+c). \end{aligned}$$

Hence the sum is divisible by 11 if $a+d$ is divisible by 11. Since a,d are from 0-9 and $a \neq 0$, this means $a+d = 11$ (8 choices). b,c can be any of $0,1,\ldots,9$ (10 choices, each) for a total of $8 \cdot 10 \cdot 10 \cdot 1 = 800$ numbers.

Answer: 800

Problem 7 Solution

A dodecagon has twelve sides, so we can divide the inscribed dodecagon into 12 isosceles triangles, as shown below.

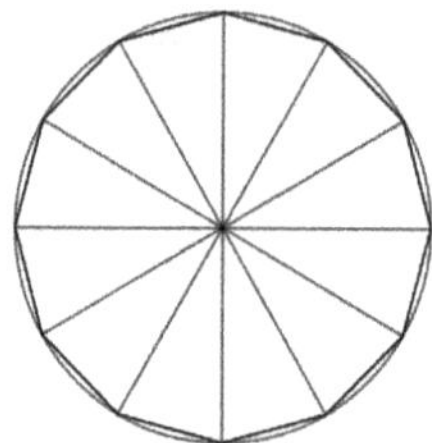

Each of these isosceles triangles has small angle of $360^\circ \div 12 = 30^\circ$, so the remaining two angles are each 75°. Drawing an altitude from one of these angles gives a 30-60-90 triangle with hypotenuse 4. Hence the height of the triangle is 2 so each of these triangles has area $\frac{1}{2} \cdot 4 \cdot 2 = 4$. This gives the total area of the dodecagon as $4 \times 12 = 48$.

Answer: 48

Problem 8 Solution

There are 5^5 ways for the 5 people to choose their favorite books. For the median of the books to be the last, we need at least 3 people to choose the last book as their favorite. For the case where exactly 3 people choose the last book as their favorite, there are $\binom{5}{3} \cdot 4^2$ outcomes, as we need to decide which 3 chose the last as their favorite and which of the other books was the favorite for the other two. Identical reasoning gives $\binom{5}{4} \cdot 4^1$ and $\binom{5}{5} \cdot 4^0$ outcomes for 4 or 5 people choosing the last book as their favorite. Hence the probability is

$$\frac{10 \cdot 16 + 5 \cdot 4 + 1}{5^5} = \frac{181}{3125}$$

so $P + Q = 181 + 3125 = 3306$.

Answer: 3306

Problem 9 Solution

Completing the square on $2x^2 - 12x + 72$ gives

$$2x^2 - 12x + 72 = 2(x^2 - 6x + 9) + 54 = 2(x-3)^2 + 54.$$

Therefore, the range of $y = \sqrt{2x^2 - 12x + 72}$ is all $y \geq \sqrt{54}$. As $7 < \sqrt{54} < 8$, the range contains all integers greater than or equal to 8 (but not 7) so $K = 8$ is the smallest such integer.

Answer: 8

Problem 10 Solution

Since the pizza has 5 toppings made available and each pizza requires 3 nonrepeated toppings, there are

$$\binom{5}{3} = 10$$

combinations of toppings that are made available for pizzas. Of the 10 possible pizzas, we wish to choose 2 pizzas for our order (without order), which can be done in

$$\binom{10}{2} = 45$$

ways.

Answer: 45

Problem 11 Solution

$\overline{AF}$, $\overline{AH}$, and $\overline{FH}$ are all face diagonals and thus

$$AF = AH = FH = \sqrt{8}$$

and the area of equilateral triangle AFH is

$$\frac{(\sqrt{8})^2\sqrt{3}}{4} = 2\sqrt{3}.$$

The surface area we want consists of $\triangle AFH$ with area $2\sqrt{3}$, 3 congruent triangles that all have area 2, and 3 squares each with area 4. Therefore its surface area is

$$2\sqrt{3}+3\cdot 2+3\cdot 4 = 18+2\sqrt{3}.$$

Hence $A+B+C = 18+2+3 = 23$.

Answer: 23

Problem 12 Solution

Divide the set $\{1,2,\ldots,500\}$ into two categories: the first category consists of numbers that are multiples of 5, and the second category consists of all other numbers. The first category contains $500 \div 5 = 100$ numbers. Two numbers have common divisor 5 only when they are both in the first category. There are $500-100 = 400$ numbers in the second category, so if we pick $400+2 = 402$ numbers, we're guaranteed to have at least two numbers with common divisor 5.

Answer: 402

Problem 13 Solution

If $y = 2x - \dfrac{1}{x}$ we have $y^2 = 4x^2 - 4 + \frac{1}{x^2}$ so substituting we have

$$y^2+2y-3 = 0 \Rightarrow (y+3)(y-1) = 0 \Rightarrow y = -3, 1.$$

If $y = -3$ we have $2x - \frac{1}{x} = -3$ or $2x^2 + 3x - 1 = 0$. Similarly if $y = 1$ we have $2x^2 - x - 1 = 0$. Solving these equations we have

$$x = -\frac{3}{4} \pm \frac{\sqrt{17}}{4}, x = -\frac{1}{2}, x = 1.$$

Hence the sum of the irrational solutions is $-\frac{6}{4} = -1.5$.

Answer: -1.5

Problem 14 Solution

Denote the heads by H and the tails by T. The order of the four H's does not matter, so put them in a line. There are at least 2 T's in between each pair of H's, so set aside $3 \times 2 = 6$ T's. The remaining 4 T's can be put in any of the 5 spaces (between the H's or at the front or back). Using stars and bars we have $\binom{4+5-1}{4} = 70$ different outcomes that fit the description.

Answer: 70

Problem 15 Solution

Using symmetry we look at spheres with centers across the diagonal:

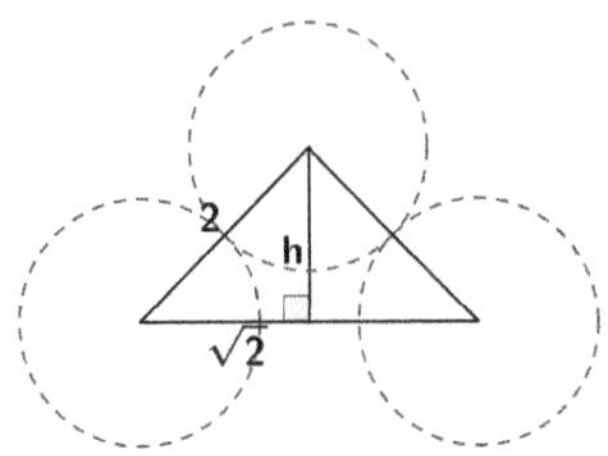

Let $h + 1$ denote the height of the center of the fifth sphere off the ground. Setting up right triangles gives us $2^2 = 2 + h^2$, so solving for h gives $\sqrt{2} \approx 1.4$.

Answer: 1.4

Problem 16 Solution

Since A has 9 divisors, it must be of the form p^8 or $p^2 \cdot q^2$ for primes p, q. Since

$$\operatorname{lcm}(A, B) = 3^2 \cdot 5 \cdot 7^2,$$

we must have

$$A = 3^2 \cdot 7^2 = 441.$$

Since $5 \nmid A$ it must be the case that $5 \mid B$, and since $\gcd(A, B) = 7$, we have $7 \mid B$. Since B has 4 factors, this implies that

$$B = 5 \cdot 7 = 35.$$

Thus the final answer is

$$A + B = 441 + 35 = 476.$$

Answer: 476

Problem 17 Solution

Let $t = \left\lfloor \frac{2x-1}{3} \right\rfloor$, so t is an integer. Then $x = \frac{2t+1}{3}$. By definition we have $t \le \frac{2x-1}{3} < t+1$, thus $t \le \frac{4t-1}{9} < t+1$. Solving for t, $0 \le -\frac{5t+1}{9} < 1$, which gives $-2 < t \le -\frac{1}{5}$. So $t = -1$. Finally, plug $t = 1$ back in to get $x = -\frac{1}{3}$.

Answer: -0.33

Problem 18 Solution

Apply the multinomial theorem to $(a+b+c)^9$ with $a=x^3$, $b=x^2$, $c=-1$. We will have terms of x^{11} of the form $a^3b^1c^5$ and $a^1b^4c^4$ (as $11=3+3+3+2$ or $11=3+2+2+2+2$). Therefore the coefficient is

$$\frac{9!}{3!5!}(-1)^5+\frac{9!}{4!4!}=\frac{9!}{3!4!}\left(\frac{1}{4}-\frac{1}{5}\right)=\frac{9!}{3!4!20}$$

which simplifies to $9\cdot7\cdot2=126$.

Answer: 126

Problem 19 Solution

Using Vieta's theorem, the roots x_1, x_2 satisfy

$$x_1+x_2=-a-14 \text{ and } x_1x_2=-a,$$

so

$$x_1x_2-x_1-x_2=14.$$

Completing the rectangle we have

$$x_1x_2-x_1-x_2+1=(x_1-1)(x_2-1)=15.$$

The possible values for the roots are then $2,16$; $0,-14$; $4,6$; and $-2,-4$. Hence a can be $-32, 0, -24, -8$. The sum of all possible values of a is -64.

Answer: -64

Problem 20 Solution

The altitudes are in ratio $12:15:20$ so the ratio of the sides is $\frac{1}{12}:\frac{1}{15}:\frac{1}{20}=5:4:3$. Therefore the triangle is right with sides $3t, 4t, 5t$ for some t and area $6t^2=96$ so $t^2=16$ and $t=4$. Hence the triangle is a $12, 16, 20$ triangle with perimeter $48=\sqrt{2304}$. Therefore $K=2304$.

Answer: 2304

2.6 ZIML March 2018 Junior Varsity

Below are the solutions from the Junior Varsity ZIML Competition held in March 2018.
The problems from the contest are available on p.45.

Problem 1 Solution

It can be determined immediately that $e = 5$. Also we know that b, d must be even digits, and so a, c are odd. So either $a = 7, c = 9$ or $c = 7, a = 9$. To make $\overline{abc}$ a multiple of 3, only $b = 8$ works. This means $d = 6$. Checking, both 78965 and 98765 will work. Thus 78965 is the smallest that works.

Answer: 78965

Problem 2 Solution

First note that $EFGH$ is in fact a kite (with $\overline{GE} \perp \overline{FH}$) and hence has area $\dfrac{GE \cdot FH}{2}$. Let the square have side length 2, so $GE = 1$.

Considering triangles $\triangle ADH$ and $\triangle GEH$ we have $\triangle ADG \sim \triangle GEH$ with ratio of sides $2:1$. This implies their heights are also in ratio $2:1$, so the height of $\triangle GEH$ is $\dfrac{1}{3} \cdot 1 = \dfrac{1}{3}$. This implies that $HF = 2 \cdot \dfrac{1}{3} = \dfrac{2}{3}$. Therefore, the area of $EFGH$ is $1 \cdot \dfrac{2}{3} \div 2 = \dfrac{1}{3}$. This is

$$\frac{1}{3} \div 2^2 = \frac{1}{12} = 0.8\overline{3} \approx 8.3\%$$

so $P = 8.3$.

Answer: 8.3

Problem 3 Solution

There are $\binom{10}{5} \cdot \binom{5}{5}$ ways to divide themselves into a first team and a second team. We do not care about the order of the teams, so we also divide by 2. This gives a final answer of $\binom{10}{5} \div 2 = 126$ total ways to divide themselves into teams.

Answer: 126

Problem 4 Solution

To avoid solving a cubic equation, write the equation as a quadratic in y:

$$0 = y^2 + (-x^2 + x - 1)y + (-x^3 + x^2) = (y - x^2)(y + x - 1).$$

Hence $y = x^2$ or $y = 1 - x$. If $y = 5$ we have $y = \pm\sqrt{5}$ or $y = -4$. The sum of all negative x-values is $-4 - \sqrt{5} \approx -6.2$.

Answer: -6.2

Problem 5 Solution

For either size we have 2 choices for which rice to purchase. Since repeated portions of seafood are allowed, we use stars and bars to count the outcomes. This gives

$$2\left(\binom{3+5-1}{3} + \binom{4+5-1}{4}\right) = 2(35+70) = 210$$

total Poke dishes.

Answer: 210

Problem 6 Solution

Let x be the shorter side, so $2x$ is the longer side and thus $x\sqrt{5}$ is the hypotenuse. Divide the unit square into four triangles (with respective bases and heights of x and $2x$, 1 and $1-2x$, 1 and $1-x$, and $x\sqrt{5}$ and 1) as shown below.

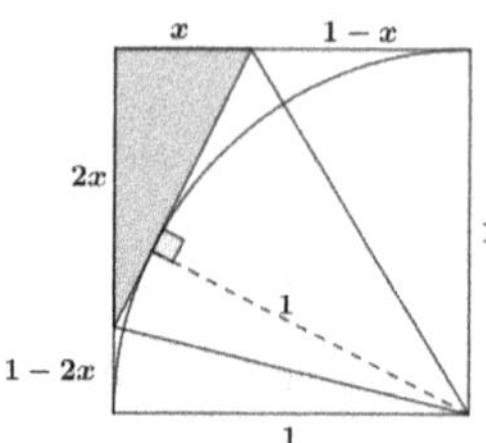

As the full square has area 1 we get

$$1 = \frac{2x^2}{2} + \frac{1-2x}{2} + \frac{1-x}{2} + \frac{x\sqrt{5}}{2}.$$

Simplifying we have $x\left(x - \frac{3}{2} + \frac{\sqrt{5}}{2}\right) = 0$ so $x = 0$ (throw out) or $x = \frac{3}{2} - \frac{\sqrt{5}}{2} = 1.5 - 0.5\sqrt{5}$. Thus

$$A + B + C = 1.5 - 0.5 + 5 = 6.$$

Answer: 6

Problem 7 Solution

Rewriting we have

$$\frac{x^4 + 4x^3 + 2x^2 - 4x + 5}{x^2 + 2x - 1} = 4.$$

Dividing the numerator by the denominator using long division we have a quotient of $x^2 + 2x - 1$ and a remainder of 4, so the equation becomes:

$$x^2 + 2x - 1 + \frac{4}{x^2 + 2x - 1} = 4.$$

Substituting $y = x^2 + 2x - 1$ we can solve to get $y = 2$. Thus, $x^2 + 2x - 1 = 2$ so $x^2 + 2x - 3 = 0$, and $x = -3$ or $x = 1$. The smallest solution is thus $x = -3$.

Answer: -3

Problem 8 Solution

We notice that $9^2 = 81$. Therefore

$$a_{n+2} = 9a_{n+1} - 1 = 9(9a_n - 1) - 1 = 81a_n - 10$$

so $a_{n+2} \equiv a_n - 10 \pmod{80}$. As

$$2018 = 2 \cdot 1009 \text{ and } 1009 \equiv 49 \pmod{80}$$

we have

$$a_{2018} \equiv a_0 - 10 \cdot 1009 \equiv 3 - 490 \equiv -487 \equiv 73 \pmod{80},$$

so the remainder is 73.

Answer: 73

Problem 9 Solution

Call the expression M, so factoring we have

$$M = \frac{(xy)^3}{(x-2y)^3} \Rightarrow \frac{1}{M} = \left(\frac{x-2y}{xy}\right)^3 = \left(\frac{1}{y} - \frac{2}{x}\right)^3.$$

Therefore $\frac{1}{M} = ((11+2\sqrt{3}) - 2(2+\sqrt{3}))^3 = 7^3$ and hence

$$M = \frac{1}{343},$$

thus $P + Q = 1 + 343 = 344$.

Answer: 344

Problem 10 Solution

There are 6^3 total outcomes. With three rolls, Justine can form a number of the form $\overline{abc}$ where each of a,b,c is chosen from $1,2,3,4,5,6$. For the number to be divisible by 11 we must have $a-b+c$ a multiple of 11. From the restrictions on a,b,c, it is clear either $a-b+c=11$ or $a-b+c=0$. The only tuple (a,b,c) with $a-b+c=11$ is $(6,1,6)$. For $a-b+c=0$ we have $a+c=b$. Note a,c are interchangeable, so for now assume $a \leq c$, giving tuples

$$(1,2,1),(2,3,1),(3,4,1),(2,4,2),(4,5,1),$$

$$(3,5,2),(5,6,1),(4,6,2),(3,6,3).$$

Including the tuple $(6,1,6)$ we have 4 tuples with repeated numbers (each can be arranged 3 ways) and 6 tuples with all distinct numbers (each can be arranged $3!=6$ ways). Hence the probability is

$$\frac{4\cdot 3+6\cdot 6}{6^3}=\frac{48}{216}=\frac{2}{9}$$

so $P+Q=2+9=11$.

Answer: 11

Problem 11 Solution

We can calculate the number of factors from the prime factorizations. We first focus on finding L. Clearly smaller primes are better. Note $2\cdot 3\cdot 5\cdot 7>100$, but $2^2\cdot 3\cdot 5=60$ has $3\cdot 2\cdot 2=12$ factors, so $L\geq 12$.

If a number has a prime factorization $2^a\cdot 3^b\cdot 5^c\cdot 7^d\cdot 11^e\cdots$, then it has $(a+1)(b+1)(c+1)\cdots$ factors. Bounding each exponent separately we have $a\leq 6$, $b\leq 4$, $c,d\leq 2$ and $e,f,\ldots,\leq 1$. Some trial and error convinces us that ≥ 13 factors is impossible, so we search for numbers with 12 factors.

$$\begin{aligned} 12 &= (11+1) &&= (5+1)(1+1) \\ &= (3+1)(2+1) &&= (2+1)(1+1)(1+1) \end{aligned}$$

so we look for numbers of the form $p^{11}, p^5q, p^3q^2, p^2qr$ (for p,q,r primes). p^{11} is impossible. For p^5q we have $2^5 \cdot 3 = 96$ (no other primes work). For p^3q^2 we again have only $p = 2, q = 3$ giving $2^3 \cdot 3^2 = 72$. For p^2qr we have $2^2 \cdot 3 \cdot 5 = 60$ and $2^2 \cdot 3 \cdot 7 = 84$ when $p = 2$ and $3^2 \cdot 2 \cdot 5 = 90$ when $p = 3$ (no others are possible).

Therefore we see that $L = 12$ and $K = 5$, and hence

$$K \times L = 5 \times 12 = 60.$$

Answer: 60

Problem 12 Solution

There are $5! = 120$ total outcomes. We use PIE (Principle of Inclusion-Exclusion) to calculate how many ways there are for at least one of them to finish in the same spot.

If we fix one of the friends to finish in the same spot, there are $\binom{5}{1}$ choices for which friend and then $4!$ ways to arrange the other friends, giving $\binom{5}{1} \cdot 4! = \frac{5!}{1!}$ outcomes. Similarly, if we fix two of the friends to finish in the same spot, there are $\binom{5}{2}$ choices for which two friends and then $3!$ ways to arrange the other friends, giving $\binom{5}{2} \cdot 3! = \frac{5!}{2!}$ outcomes. Continuing in this manner there are

$$\frac{5!}{1!} - \frac{5!}{2!} + \frac{5!}{3!} - \frac{5!}{4!} + \frac{5!}{5!}$$

ways for the friends to finish in the same spot. This gives

$$5! - \frac{5!}{1!} + \frac{5!}{2!} - \frac{5!}{3!} + \frac{5!}{4!} - \frac{5!}{5!} = 60 - 20 + 5 - 1 = 44$$

outcomes where none of the five finish in the same spot.

Answer: 44

Problem 13 Solution

The diagram is as in the diagram below.

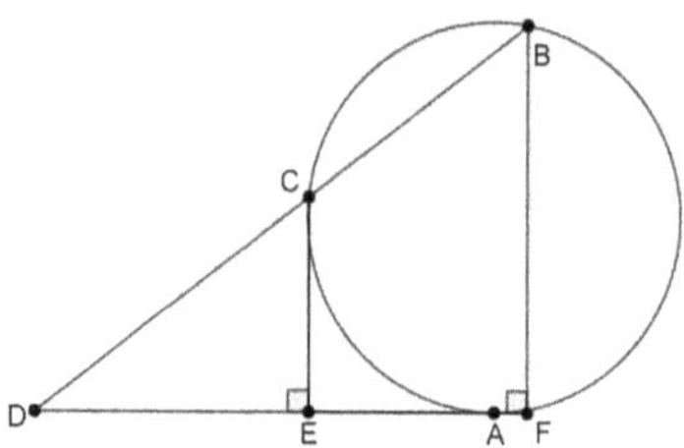

Let $CD = x$. Using Power of a Point we have

$$\begin{aligned} AD^2 = CD \cdot BD \quad &\Rightarrow \quad (6\sqrt{5})^2 = x(x+8) \\ &\Rightarrow \quad x^2 + 8x - 180 = (x+18)(x-10) = 0 \end{aligned}$$

so $CD = x = 10$ (as $x = -18$ does not work). Therefore $DE = 8$ (using Pythagorean triples) so the area of $\triangle CED$ is 24. $\triangle CED \sim \triangle BFD$ with a ratio of sides of $10 : 18 = 5 : 9$ so their areas are in ratio $25 : 81$ and $\triangle BFD$ has area $\frac{81}{25} \cdot 24$. Therefore the area of $BCEF$ is

$$\frac{81}{25} \cdot 24 - 24 = \frac{56}{25} \cdot 24 = \frac{1344}{25} = 53.76$$

so the area rounded to the nearest integer is 54.

Answer: 54

Problem 14 Solution

The prime factorization of $10!$ is $2^8 \cdot 3^4 \cdot 5^2 \cdot 7$. Hence $\sqrt[4]{10!} = 2^2 \cdot 3 \cdot \sqrt[4]{5^2 \cdot 7}$. Therefore $\sqrt[4]{10!}$ is between $12 \cdot 3 = 36$ and $12 \cdot 4 = 48$.

The only multiple of 7 in this range is 42, so we try to set one number to $42 = 2 \cdot 3 \cdot 7$. Similarly for the factors of 5 we try $40 = 2^3 \cdot 5$ and $45 = 3^2 \cdot 5$. We are left with $2^4 \cdot 3 = 48$. In this case $A, B, C, D = 40, 42, 45, 48$ with $D - A = 8$.

No other rearrangement of the prime factors leads to a smaller value for $D - A$, leaving 8 as our answer.

Answer: 8

Problem 15 Solution

The shortest path is a straight line in the unfolded cube as shown below.

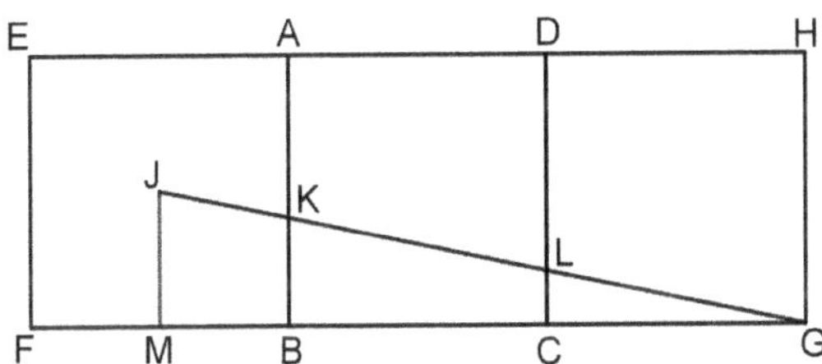

We know that $JM = 4 \div 2 = 2$ and $GM = 4 + 4 + 2 = 10$ and hence $JG = \sqrt{2^2 + 10^2} = \sqrt{104}$. $\triangle JMG \sim \triangle KBG \sim \triangle LCG$ with ratio of sides $5 : 4 : 2$ so it follows that $KL : JG = 2 : 5$ and hence $KL = \frac{2}{5} \cdot \sqrt{104}$. As $10^2 < 104 < 10.5^2$ we know $10 < \sqrt{104} < 10.5$ and hence $4 < KL < 4.2$ so $KL \approx 4$.

Answer: 4

Problem 16 Solution

Let r,s be the roots of the quadratic with $|r| < |s|$. Using Vieta's formulas we have $r+s=-a$ and $rs=1-b$. Note this implies that a,b are also integers. We have

$$\begin{aligned} 10 &= a^2+b^2 \\ &= (r+s)^2+(rs-1)^2 \\ &= r^2+2rs+s^2+r^2s^2-2rs+1 \\ &= (r^2+1)(s^2+1). \end{aligned}$$

As $10=1\cdot 10=2\cdot 5$ we have $r=0, s=\pm 3$ or $r=\pm 1, s=\pm 2$ (6 total possibilities). Checking, each pair of (r,s) gives a unique pair (a,b), so there are 6 pairs for (a,b). (These pairs are $(a,b)=(\pm 3,\pm 1),(\pm 1,3)$.)

Answer: 6

Problem 17 Solution

Without loss of generality suppose they tied by both picking a card of value 1 the first time. For the second game, the deck has 23 total cards, 3 with value 1 and 5 each for the values $2,3,4,5$.

To tie again Kate and then Leo could each pick a card of value 1 again, which has probability $\frac{3}{23}\cdot\frac{2}{22}$. Otherwise Kate can pick one of the other 20 cards, and there are 4 remaining for Leo to pick so that they tie. This has probability $\frac{20}{23}\cdot\frac{4}{22}$. Therefore the probability is

$$\frac{3}{23}\cdot\frac{2}{22}+\frac{20}{23}\cdot\frac{4}{22}=\frac{86}{506}=\frac{43}{253}$$

and $Q-P=253-43=210$.

Answer: 210

Problem 18 Solution

Based on the equation, there are two possibilities:

$$(1)\ x^2+ax+1=x^2+x+a;\ (2)\ x^2+ax+1=-(x^2+x+a).$$

Consider Case (1). If $a=1$, both sides are exactly the same, so there are infinitely many solutions, not what we want. If $a\neq 1$, we get

$$ax+1=x+a,$$

so

$$x=\frac{a-1}{a-1}=1.$$

For Case (2), we have

$$2x^2+(a+1)x+(a+1)=0.$$

As $x=1$ is already one solution from Case (1), there are two possibilities for this quadratic equation. First possibility: this quadratic equation has exactly one real root. This means the discriminant is 0. Thus

$$\begin{aligned} & (a+1)^2-4\cdot 2\cdot (a+1)=0 \\ \Rightarrow\ & (a+1)(a+1-8)=0 \\ \Rightarrow\ & a=-1 \text{ or } a=7. \end{aligned}$$

The other possibility is that $2x^2+(a+1)x+(a+1)=0$ has two real roots, but one of them is $x=1$ (leading to two solutions in total). This happens when $2+(a+1)+(a+1)=0$, or $a=-2$.

Combining, the equation

$$|x^2+ax+1|=|x^2+x+a|$$

has exactly two real solutions when $a=-2$, $a=-1$, or $a=7$. The sum of these values is $-2-1+7=4$.

Answer: 4

Problem 19 Solution

Let O be the center of C_1 and P the center of C_2. We have a diagram similar to the one found below.

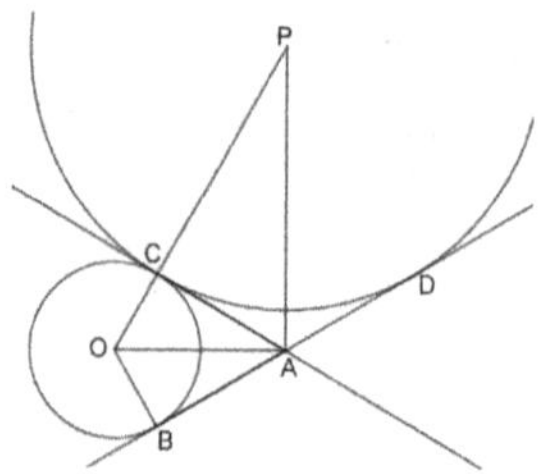

As $\angle BAC = 60^\circ$ we have that $\triangle ABO$ and $\triangle ACO$ are congruent 30-60-90 triangles. As $OB = OC = 1$ we have $OA = 2$. An identical argument gives $\angle PAC = \angle CAD \div 2 = 120^\circ \div 2 = 60^\circ$ so in face $\triangle OAP$ is a 30-60-90 triangle as well. Therefore $OP = 2 \cdot OA = 4$ which gives $PC = OP - OC = 4 - 1 = 3$. Thus C_2 has a radius of length 3.

Answer: 3

Problem 20 Solution

Clearly $x \geq 0$. Let $t = \lfloor 4x \rfloor$ (so $t \geq 0$ is an integer). Then $x^2 = t$ so $x = \sqrt{t}$. We know that

$$t \leq 4x < t+1 \Rightarrow t \leq 4\sqrt{t} < t+1 \Rightarrow t^2 \leq 16t < t^2 + 2t + 1.$$

Solving these inequalities we have (recall $t \geq 0$) we get (i) $t \leq 16$ and (ii) $t^2 - 14t + 1 > 0$. Using the quadratic formula for

$$t^2 - 14t + 1 = 0$$

we get roots $7 \pm 4\sqrt{3}$ so

$$t \leq 7 - 4\sqrt{3} \approx 0.07 \text{ or } t \geq 7 + 4\sqrt{3} \approx 13.93.$$

As t is an integer, we have $t = 0, 14, 15, 16$. Hence the equation has solutions $\sqrt{0}, \sqrt{14}, \sqrt{15}, \sqrt{16}$ so the sum of all K is

$$0 + 14 + 15 + 16 = 45.$$

Answer: 45

2.7 ZIML April 2018 Junior Varsity

Below are the solutions from the Junior Varsity ZIML Competition held in April 2018.
The problems from the contest are available on p.53.

Problem 1 Solution

For the last two digits we work $\pmod{100}$. Note that

$$7^4 \equiv 2401 \equiv 1 \pmod{100}.$$

Thus we need to calculate $7^7 \pmod{4}$. We have

$$7^7 \equiv (-1)^7 \equiv -1 \equiv 3 \pmod{4}.$$

Therefore

$$7^{7^7} \equiv 7^3 \equiv 343 \equiv 43 \pmod{100}$$

so the last two digits are 43.

Answer: 43

Problem 2 Solution

If we expand $(x-1)^7$ we would get

$$(x-1)^7 = Ax^7 + Bx^6 + Cx^5 + \cdots + Gx + H.$$

Note then if we set $x = 1$ we have

$$(1-1)^7 = A(1)^7 + B(1)^6 + C(1)^5 + \cdots + G(1) + H,$$

so in fact

$$0 = A + B + C + \cdots + G + H$$

and the sum of all the coefficients of this polynomial is 0.

Of course you can use Pascal's triangle to help actually calculate each of the coefficients, but this is an efficient solution.

Answer: 0

Problem 3 Solution

On each of the 9 floors the elevator either stops or does not stop, 2 choices. This gives 2^9 total outcomes. However, the elevator must stop on at least one floor (so we must remove 1 outcome). Further, because we have 8 people the elevator cannot stop on all 9 floors (so we must remove 1 more outcome). Hence there are $2^9-2=512-2=510$ different collections of floors the elevator can stop on.

Answer: 510

Problem 4 Solution

Using isosceles triangles, it is easy to see that $\angle BPC = \angle BCP = 75^\circ$. Since $\angle BCA = 45^\circ$, we get $\angle PCQ = 75^\circ - 45^\circ = 30^\circ$. Hence $\angle PQC = 75^\circ = \angle CPQ$. So $\triangle CPQ$ is a 30°-75°-75° triangle. Hence the largest angle minus the smallest angle is $75^\circ - 30^\circ = 45^\circ$.

Answer: 45

Problem 5 Solution

The number must be divisible by 3 and by 5. To be divisible by 5 the last number must be 5. To be divisible by 3, the sum of the 5 digits must be divisible by 3. Since $\gcd(3,5)=1$ this means the digit 5 must appear 3 times in total (so the digit 3 appears twice).

The last digit is 5, so the other four digits consist of two 3's and two 5's, hence there are 6 possible numbers: 33555, 35355, 35535, 53355, 53535, and 55335. To easily add up these numbers, note that in the ones place the digit 5 appears all 6 times. In the other places (tens, hundreds, etc.) each digit appears 3 times. Hence the sum is

$$\begin{aligned} & (3\cdot 3+3\cdot 5)\cdot(10^4+10^3+10^2+10)+(6\cdot 5)\cdot 1 \\ = \ & 24\cdot(10^4+10^3+10^2+10)+30 \\ = \ & 266670 \end{aligned}$$

as our final answer.

Answer: 266670

Problem 6 Solution

Let $x = \sqrt{2018}$. Then we have

$$\begin{aligned} L &= \frac{x^3(x+4) - 3x(x-4) + 1}{x^2+4x} \\ &= \frac{x^4 + 4x^3 - 3x^2 + 12x + 1}{x^2+4x} \\ &= x^2 - 3 + \frac{24x+1}{x^2+4x} \end{aligned}$$

using polynomial long division. Hence as $\frac{24x+1}{x^2+4x} < 1$ (see the details below), we have $\lfloor L \rfloor = x^2 - 3 = 2018 - 3 = 2015$.

One possible argument that $\frac{24x+1}{x^2+4x} < 1$ is as follows. Note $40 < x < 50$ so

$$\frac{24x+1}{x^2+4x} < \frac{24 \cdot 50 + 1}{x^2+4x} < \frac{1201}{40^2} = \frac{1201}{1600} < 1,$$

as claimed.

Answer: 2015

Problem 7 Solution

First, the planes containing each of the 6 faces clearly work, giving 6 different planes.

Consider one edge of the cube. The plane containing this edge and the opposite diagonal edge (parallel to the first edge so the plane cuts the cube in half) also contains exactly 4 vertices. There are $12 \div 2 = 6$ of these planes as well.

In fact these are the only two types of planes that work, so there are $6+6=12$ total planes that pass through exactly 4 vertices of the cube.

Answer: 12

Problem 8 Solution

The full sum is $45 \cdot 14 \div 2 = 315$. Note that any sum 0, 3, 6, ..., 315 is thus possible. Thus there are $315 \div 3 + 1 = 106$ possible sums.

Answer: 106

Problem 9 Solution

Every number has a prime factorization. The smallest primes larger than 29 are $31, 37, 41 \ldots$. Thus the smallest possible composite numbers not deleted are $31^2, 31 \cdot 37,$, etc. The squares of the form p^2 all have exactly 3 factors, so the smallest with 4 or more factors is $31 \cdot 37 = 1147$.

Answer: 1147

Problem 10 Solution

$\triangle ABM$ and $\triangle ADM$ share the same height (from A to $\overline{BD}$) so $BM : DM = [ABM] : [ADM]$ (here $[ABM]$ and $[ADM]$ denote the area of each triangle). An identical argument also gives $BM : DM = [CBM] : [CDM]$. Therefore

$$\begin{aligned} BM : DM &= [ABM] + [CBM] : [ADM] + [CDM] \\ &= [ABC] : [ADC] \\ &= 10 : 6 = 5 : 3, \end{aligned}$$

and hence (as $BM + DM = 8$) $BM = 5$.

Answer: 5

Problem 11 Solution

20 is even, so the only way to write $20 = 2 \cdot A + 3 \cdot B$ (with A, B non-negative integers) is when $(A,B) = (10,0)$, $(7,2)$, $(4,4)$, or $(2,6)$, so we consider 4 cases.

In the first case, the list is just ten 2's, which can only be arranged in 1 way. In the second case, the list is seven 2's and two 3's. One 2 must be reserved to separate the 3's. Then arrange the other six 2's into $2+1=3$ spots using stars and bars. This gives $\binom{6+3-1}{6} = 28$ additional lists. In the third case, the list is four 2's and four 3's. Three of the 2's must be reserved to separate the 3's, so we only have $4+1=5$ spots for the last 2. The fourth case is impossible, as we do now have enough 2's to separate the 3's.

Hence there are $1+28+5=34$ total lists George could have written during the basketball game.

Answer: 34

Problem 12 Solution

Let $u = x+y$ and $v = xy$. Then $x^2+y^2 = u^2-2v$. Substituting into the left-hand side of the equation we have

$$u^4 - 6v(u^2-2v) - 4v^2 = u^4 - 6u^2v + 8v^2 = (u^2-4v)(u^2-2v),$$

so plugging back u, v we get $((x+y)^2-4xy)((x+y)^2-2xy) = (x^2-2xy+y^2)(x^2+y^2) = (x-y)^2(x^2+y^2)$. As x, y are integers, we must either have $(x-y)^2 = 1$ or $(x-y)^2 = 4$ (since 1 and 4 are the only squares that are factors of 40).

In the first case, $|x-y| = 1$ and $x^2+y^2 = 40$, but no such pairs (x,y) exist over the integers. In the second case, we need $|x-y| = 2$ and $x^2+y^2 = 10$. The only way to express 10 as the sum of two squares is 1^2+3^2, giving the pairs $(3,1)$, $(1,3)$, $(-3,-1)$, and $(-1,-3)$. Hence there are 4 pairs in total.

Answer: 4

Problem 13 Solution

Recall that the perpendicular bisector of two points contains all the points that are equal distance from both points. Using the perpendicular bisectors we divide the square into the three regions shown on the left below, with each region closest to A, B, or C.

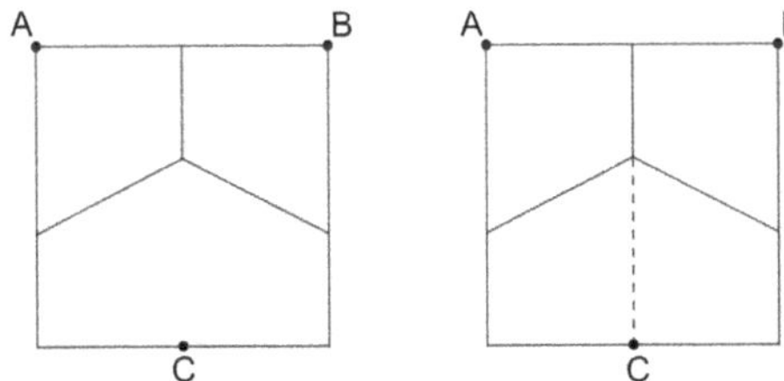

Extending the vertical line, as shown on the right above, divides the square into 4 congruent trapezoids. Hence the probability a randomly chosen point inside the square is closest to A is 25% and $K = 25$.

Answer: 25

Problem 14 Solution

$\triangle ABC$ is a 30-60-90 triangle and thus $BC = 2\sqrt{3}$ and $AC = 3$. Quadrilateral $ABED$ is a cyclic quadrilateral, and hence $\angle BED = 180° - \angle BAD = 90°$ and $\angle ADE = 180° - \angle ABE = 120°$. Therefore $\triangle CED$ is also a 30-60-90 right triangle. As $\angle BAD = 90°$, $\overline{BD}$ is a diameter of the circle, hence $BD = 2$. Thus $AD = \sqrt{2^2 - (\sqrt{3})^2} = 1$ and $CD = AC - AD = 3 - 1 = 2$.

Thus $ED = 1$, $EC = \sqrt{3}$ so $\triangle CED$ has area $\dfrac{\sqrt{3}}{2}$. As $\triangle ABC$ has area $\frac{3\sqrt{3}}{2}$, quadrilateral $ABED$ has area $\sqrt{3} \approx 1.73$.

Answer: 1.7

Problem 15 Solution

We work modulo 3. We want $n \cdot 2^n \equiv 2 \pmod 3$.

First consider $n \equiv 2 \pmod 3$ and $2^n \equiv 1 \pmod 3$. As $2^2 \equiv 4 \equiv 1 \pmod 3$, $2^n \equiv 1 \pmod 3$ when n is even. Hence all $n \equiv 2 \pmod 6$ work.

Next consider $n \equiv 1 \pmod 3$ with $2^n \equiv 2 \pmod 3$. We have $2^2 \equiv 2 \pmod 3$ when n is odd. Hence all $n \equiv 1 \pmod 6$ also work.

Hence $a_n \equiv 0 \pmod 3$ when either $n = 1 + 6k$ or $n = 2 + 6k$ for integers k. As $6 \cdot 16 = 96$, $k = 0, \ldots, 16$ all work. This means there are $17 \cdot 2 = 34$ total n such that a_n has remainder 0 when divided by 3.

Answer: 34

Problem 16 Solution

Using Vieta' theorem we have

$$abcd = \frac{5}{2} \text{ and } abc + abd + acd + bcd = -\frac{-4}{2} = 2$$

Therefore, with a common denominator we have

$$\frac{1}{a} + \frac{1}{b} + \frac{1}{c} + \frac{1}{d} = \frac{bcd + acd + abd + abc}{abcd} = 2 \div \frac{5}{2} = \frac{4}{5}.$$

Thus $P - Q = 4 - 5 = -1$.

Answer: -1

Problem 17 Solution

Let $AE = x$ and $AF = y$. Hence $EF = 2 - x - y$ and we want to find $\frac{xy}{2}$. Using the Pythagorean theorem, $AE^2 + AF^2 = EF^2$ so

$$x^2 + y^2 = (2 - x - y)^2 \Rightarrow 0 = 2xy - 4x - 4y + 4 \Rightarrow x + y = \frac{xy}{2} + 1.$$

As the area of $\triangle CEF$ is the full square minus the three outside triangles we also have

$$\frac{17}{40} = 1 - \frac{xy}{2} - \frac{1-x}{2} - \frac{1-y}{2} = \frac{x+y}{2} - \frac{xy}{2}.$$

Substituting $x + y = \frac{xy}{2} + 1$ we get

$$\frac{17}{40} = \frac{1}{2} - \frac{xy}{4} \Rightarrow \frac{xy}{2} = 2 \cdot \left(\frac{1}{2} - \frac{17}{40}\right) = \frac{3}{20}.$$

This is the area of $\triangle AEF$ so $P + Q = 3 + 20 = 23$.

Answer: 23

Problem 18 Solution

Use $x^4 = -2x + 17$ to reduce the exponents.

$$\begin{aligned} x^7 + 2x^4 - 17x^3 &= x^4(x^3 + 2) - 17x^3 \\ &= (-2x + 17)(x^3 + 2) - 17x^3 \\ &= -2x^4 + 17x^3 - 4x + 34 - 17x^3 \\ &= -2(x^4 + 2x) + 34 \\ &= -2(17) + 34 = 0 \end{aligned}$$

Hence $x^7 + 2x^4 - 17x^3 = 0$ for x satisfying $x^4 + 2x = 17$.

Answer: 0

Problem 19 Solution

We want the probability that 3 of the slips at the end are green, so when we first choose three strips, we need 2 or 3 of them to be green (so 4 or 6 are green when we cut them in half). We know from the original 5 slips there are $\binom{5}{3} = 10$ total outcomes.

$\binom{3}{3} = 1$ of these outcomes results in all 3 green slips being chosen. In this case the probability of choosing 3 green slips after they are cut in half is 1.

$\binom{3}{2}\binom{2}{1} = 6$ of these outcomes result in 2 green and 1 red slip being chosen. Hence there are 4 green and 2 red slips to choose from at the end. In this case the probability is

$$\binom{4}{3}\binom{2}{2} \div \binom{6}{3} = \frac{4}{20} = \frac{1}{5}$$

that the 3 slips are green.

Using the law of total probability the final probability is

$$\frac{1}{10} \cdot 1 + \frac{6}{10} \cdot \frac{1}{5} = \frac{11}{50}$$

so $Q - P = 50 - 11 = 39$.

Answer: 39

Problem 20 Solution

$100 - 5 = 95$, so the number of pirates originally was a factor of 95. $100 - 10 = 90$, so after throwing one pirate overboard, the remaining number of pirates is a factor of 90.

We know $95 = 5 \times 19$ so originally there were 1, 5, 19, or 95 pirates. 1 or 5 pirates doesn't make sense (as then there wouldn't be 5 coins left over), so after throwing one pirate overboard there must be 18 or 94 pirates. Only 18 is a factor of 90, so there must have been 19 pirates originally.

The coins can be distributed evenly when the number of pirates remaining is a factor of 100. The largest factor of 100 less than 19 is 10, and therefore when the process is completed, 10 pirates will each receive $100 \div 10 = 10$ coins.

Answer: 10

2.8 ZIML May 2018 Junior Varsity

Below are the solutions from the Junior Varsity ZIML Competition held in May 2018.
The problems from the contest are available on p.59.

Problem 1 Solution

Factoring we have $(x-3)(x-1)(x-1)(x+1)=5$. Notice if we regroup we have

$$(x-3)(x+1)(x-1)^2=(x^2-2x-3)(x^2-2x+1)=5.$$

Hence if we make the substitution $y=x^2-2x-3$ we have $y(y+4)=5$ or $y^2+4y-5=(y+5)(y-1)=0$ so $y=-5$ or $y=1$. Thus we get $x^2-2x+2=0$ or $x^2-2x-4=0$. The first equation does not have real solutions, but for $x^2-2x-4=0$ we have solutions $1\pm\sqrt{5}$ using the quadratic formula. Therefore

$$A+B=1+5=6.$$

Answer: 6

Problem 2 Solution

First note all the numbers Roger likes are also liked by Phil. Therefore we just need to consider numbers liked by at least one of Phil, Steve, or Tony. Let P, S, and T denote the three sets of numbers respectively. Using $n(P), n(S), n(T)$ to denote the size of these sets, we want

$$\begin{aligned} n(P\cup S\cup T) &= n(P)+n(S)+n(T) \\ &\quad -n(P\cap S)-n(P\cap T)-n(S\cap T) \\ &\quad +n(P\cap S\cap T) \end{aligned}$$

using the Principle of Inclusion-Exclusion. We have $n(P)=50$ as half the numbers are even, $n(S)=10$ as we have the squares

$1^2, \ldots, 10^2$, and $n(T) = 11$ as we have the multiples of 9 from 9 to 99. The even squares are $2^2, 4^2, \ldots, 10^2$ so $n(P \cap S) = 5$ while the even multiples of 9 are $18, 36, \ldots, 90$ so $n(P \cap T) = 5$ as well. The only squares that are multiples of 9 are 9 and 81, so $n(S \cap T) = 2$. Lastly, none of the numbers are even, a square, and a multiple of 9, so $n(P \cap S \cap T) = 0$. This gives

$$n(P \cup S \cup T) = 50 + 10 + 11 - 5 - 5 - 2 = 59$$

numbers liked by at least one of Phil, Roger, Steven, or Tony.

Answer: 59

Problem 3 Solution

By symmetry we only consider lines that intersect to the left of the given edge. Further, we can ignore parallel lines, so we only consider the lines extending $\overline{EF}$, $\overline{EG}$, and $\overline{EH}$ as shown below:

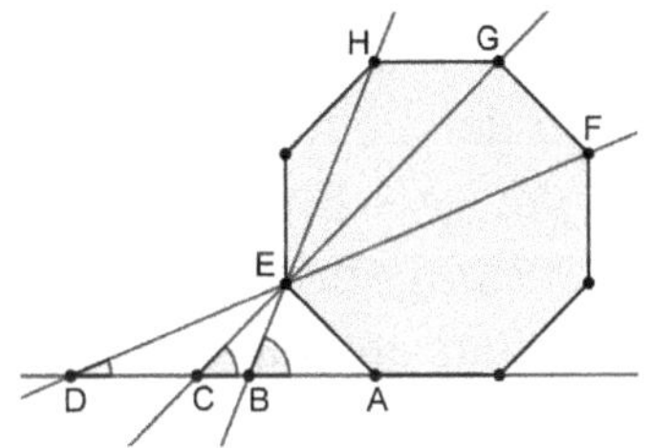

$\angle ADE$ is the smallest, so we need to find its measure. E and F are opposite vertices, so $\angle AEF$ is half the interior angle of the hexagon, hence $135^\circ \div 2 = 67.5^\circ$. Thus $\angle AED = 112.5^\circ$. $\angle DAE = 45^\circ$ as it is an external angle of the hexagon. Therefore

$$\angle ADE = 180^\circ - 112.5^\circ - 45^\circ = 22.5^\circ,$$

the smallest such angle.

Answer: 22.5

Problem 4 Solution

We want to find the sum modulo 49. First note

$$1^3 + 2^3 + \cdots + 49^3 \equiv 50^3 + 51^3 + \cdots + 98^3 \pmod{49}$$

so our sum is congruent to $2(1^3 + \cdots + 49^3) + 99^3 \pmod{49}$. $49 \equiv 0 \pmod{49}$. As $48 \equiv -1 \pmod{49}$, $47 \equiv -2 \pmod{49}$ we have

$$\begin{aligned} & 1^3 + 2^3 + \cdots + 49^3 \\ \equiv\; & 1^3 + \cdots + 24^3 + (-24)^3 + (-23)^3 + \cdots + (-1)^3 + 0^3 \\ \equiv\; & 0 \pmod{49}. \end{aligned}$$

Thus our sum is equivalent to $2 \cdot 0 + 99^3 \equiv 0 + 1^3 \equiv 1 \pmod{49}$ so the remainder is 1.

Answer: 1

Problem 5 Solution

Ania and Duncan eat 1 and 4 tamales for sure, leaving 7 more tamales. Let a, b, c, d denote the tamales Ania, Bridget, Carlos, and Duncan eat (so $a + b + c + d \leq 7$). If we let e denote the leftover tamales that Emily brings home, we then need $a + b + c + d + e = 7$. Using stars and bars there are

$$\binom{7+5-1}{7} = \binom{11}{7} = 330$$

total ways to choose a, b, c, d, e. As these also determine how many tamales each of the friends eat, 330 is our final answer.

Answer: 330

Problem 6 Solution

First we have

$$11 - |11 - |11 - 2x|| = \pm 2,$$

so

$$|11-|11-2x||=11\pm 2.$$

Repeating this process for 9 and 13 we get $|11-2x|=11\pm 9$ or $|11-2x|=11\pm 13$. Repeating one last time for $2,20,24$ (as -2 is impossible) we get

$$x=\frac{11\pm 2}{2},\frac{11\pm 20}{2},\frac{11\pm 24}{2}\Rightarrow x=\frac{13}{2},\frac{9}{2},-\frac{9}{2},\frac{31}{2},-\frac{13}{2},\frac{35}{2},$$

so the sum of the solutions is $\frac{31}{2}+\frac{35}{2}=33.$

Answer: 33

Problem 7 Solution

For A to have 8 divisors, it must be of the form p^7 or p^3q or pqr for primes p,q,r. As $\operatorname{lcm}(A,B)=6615=3^3\cdot 5\cdot 7$ the first form is impossible. For the second form both $A=3^3\cdot 5$ and $A=3^3\cdot 7$ are possible and we have $A=3\cdot 5\cdot 7$ for the final form.

If $A=3^3\cdot 5$ or $A=3^3\cdot 7$, B must be a multiple of 7^2. In fact, if $A=3^3\cdot 5$, $B=7^2$ gives $\operatorname{lcm}(A,B)=6615$, so our answer is a factor of 49. If $A=3\cdot 5\cdot 7$, B must be a multiple of $3^3\cdot 7^2$, so we see $7^2=49$ is the greatest common factor of all such integers B.

Answer: 49

Problem 8 Solution

As $AB=1$, $AC=\sqrt{3}$ so $AD=CD=\frac{\sqrt{3}}{2}$. Hence using the Pythagorean theorem $BD=\frac{\sqrt{7}}{2}$. Therefore using Power of a Point we have

$$AD\cdot CD=BD\cdot DE\Rightarrow\frac{\sqrt{3}}{2}\cdot\frac{\sqrt{3}}{2}=\frac{\sqrt{7}}{2}\cdot DE$$

Solving for DE we have $DE = \frac{3}{2\sqrt{7}} = \frac{3\sqrt{7}}{14}$ so

$$R+S+T = 3+7+14 = 24.$$

Answer: 24

Problem 9 Solution

Rewriting with a common denominator of rst we have (note we are not combining anything yet)

$$\begin{aligned}
&\frac{(r^2t+r^2s+rst)+(s^2t+rs^2+rst)+(st^2+rt^2+rst)}{rst}\\
=&\frac{r(rt+rs+st)+s(st+rs+rt)+t(st+rt+rs)}{rst}\\
=&\frac{(r+s+t)(rs+rt+st)}{rst}
\end{aligned}$$

Using Vieta's formulas we have $r+s+t = -(-4) = 4$, $rs+rt+st = 2$, and $rst = -4$. Hence the expression is $\frac{4\cdot 2}{-4} = \frac{-2}{1}$ and $P+Q = -2+1 = -1$.

Answer: -1

Problem 10 Solution

We first note that $ABED$ is a trapezoid. This implies that $[ABF] = [DEF]$ as well as $\triangle BEF \sim \triangle DAF$. As $BE = \frac{2}{3}BC = \frac{2}{3}AD$, the ratio of sides is $2:3$ so

(i) $[BED] = \frac{2}{3}\cdot\frac{1}{2}[ABCD] = 60$,
(ii) $BF : DF = 2:3 \Rightarrow [BEF] : [DEF] = 2:3$.

Thus $[DEF] = \frac{3}{5} \cdot 60 = 36$ is our final answer.

Answer: 36

Problem 11 Solution

31 is prime, so by Fermat's Little Theorem, $3^{30} \equiv 1 \pmod{31}$. As $2018 \div 30$ has remainder 8, we need to calculate $3^8 \pmod{31}$. We have $3^4 \equiv 81 \equiv -12 \pmod{31}$. Therefore

$$3^{2018} \equiv 3^8 \equiv (3^4)^2 \equiv (-12)^2 \equiv 144 \equiv 20 \pmod{31}$$

and hence the remainder is 20.

Answer: 20

Problem 12 Solution

Ignoring the card Alex discards, we need Alex to be dealt at least 4 red cards or at least 4 black cards. Hence we have 4 cases: 5 red and 0 black, 4 red and 1 black, 1 red and 4 black, or 0 red and 5 black. As there are $\binom{10}{5}$ total outcomes the probability is

$$\begin{aligned} & \frac{\binom{5}{5}\binom{5}{0} + \binom{5}{4}\binom{5}{1} + \binom{5}{1}\binom{5}{4} + \binom{5}{0}\binom{5}{5}}{\binom{10}{5}} \\ = & \frac{1+25+25+1}{252} \\ = & \frac{52}{252} \\ = & \frac{13}{63} \end{aligned}$$

so $Q - P = 63 - 13 = 50$.

Answer: 50

Problem 13 Solution

Isolating we have $\sqrt{3x-5} = \sqrt{5x+1} - 2$. Squaring we have $3x - 5 = 5x + 1 - 4\sqrt{5x+1} + 4$ or $2\sqrt{5x+1} = x + 5$. Hence squaring again we have $4(5x+1) = x^2 + 10x + 25$ so $x^2 - 10x - 21 = 0$. Factoring we get $(x-3)(x-7) = 0$ so $x = 3$ or $x = 7$. Checking, both are solutions, with sum $3 + 7 = 10$.

Answer: 10

Problem 14 Solution

Let $CD = x$. By the Pythagorean theorem

$$AD^2 = (x+5)^2 + 12^2 = x^2 + 10x + 169.$$

Using the angle bisector theorem we have

$$\frac{BC}{CD} = \frac{AB}{AD} \Rightarrow \frac{5}{x} = \frac{12}{\sqrt{x^2 + 10x + 169}} \Rightarrow 5\sqrt{x^2 + 10x + 169} = 12x.$$

Squaring both sides we have $25x^2 + 250x + 169 \cdot 25 = 144x$ or $119x^2 - 250x + 169 \cdot 25 = 0$. Noting that $169 - 119 = 50$ we can factor this as $(x+5)(119x - 845) = 0$ so solving we have $x = \frac{845}{119}$ (as $x = -5$ does not make sense). From above this gives us

$$\frac{BC}{CD} = \frac{AB}{AD} \Rightarrow AD = \frac{CD \cdot AB}{BC} = \frac{845}{119} \cdot \frac{12}{5} = \frac{2028}{119}.$$

Hence $N + M = 2028 + 119 = 2147$.

Answer: 2147

Problem 15 Solution

$2x + 1$ must be an integer, so let $t = 2x + 1$ an integer. Hence $x = \frac{t-1}{2}$. We know that

$$t \le 3x - 3 < t + 1 \Rightarrow t \le 3 \cdot \frac{t-1}{2} - 3 < t + 1.$$

Solving for t we get $9 \leq t < 11$ so $t = 9$ or $t = 10$ as t is an integer. Hence $x = 4$ or $x = 4.5$, so the largest solution is 4.5.

Answer: 4.5

Problem 16 Solution

First consider just which color he used. He used 1 red, 2 white, and 3 blue golf balls, so there are $\dfrac{6!}{1! \cdot 2! \cdot 3!} = 60$ ways to arrange just the colors. There is only one blue golf ball, but there are 2 white and 3 red, so there are a further $2^2 \cdot 3 = 12$ ways to choose which number he used each time. This gives a total of $60 \cdot 12 = 720$ total outcomes.

Answer: 720

Problem 17 Solution

Let $t = 0$ denote the time the race starts (in minutes). As 75 seconds is 1.25 minutes, Sarah gives Elizabeth a 1.25 minute head start. If we let the point (x, y) denote when Elizabeth finishes the race (the x-coordinate) and when Sarah finishes the race (the y-coordinate) these points form the rectangle where $7 \leq x \leq 10$ and $7.25 \leq y \leq 9.25$. Sarah wins the race when $y \leq x$ as shown below:

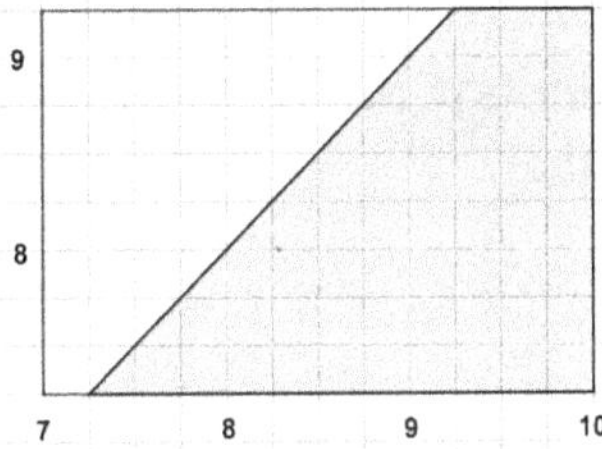

Therefore the probability Sarah wins is the ratio of the shaded region to the entire rectangle. If each box in the grid below represents 1 unit, the total area is $12 \cdot 8 = 96$. The shaded region (a trapezoid) has area $\frac{11+3}{2} \cdot 8 = 56$. Therefore the probability Sarah wins the race is $\frac{56}{96} = \frac{7}{12} \approx 0.5833$. Therefore $K = 58.\overline{3}$ so rounded to the nearest integer is 58.

Answer: 58

Problem 18 Solution

Rearranging the equation we have $x^2 + 3kx + 3k^2 - 27 = 0$. For the equation to have real roots we need the discriminant $\Delta \geq 0$. Hence

$$(3k)^2 - 4 \cdot (3k^2 - 27) \geq 0 \Rightarrow -3k^2 + 108 \geq 0 \Rightarrow k^2 \leq 36.$$

Therefore $-6 \leq k \leq 6$ if the equation has real root(s). For the roots to be non-negative, we need the sum and product of the roots to be ≥ 0. Using Vieta's formulas this means

$$-(3k) \geq 0 \text{ and } 3k^2 - 27 \geq 0 \Rightarrow k \leq 0 \text{ and } k^2 \geq 9.$$

Combining these we have $k \leq -3$. As we also know $k \geq -6$ the integer values of k that work are -6, -5, -4, and -3, a total of 4 possibilities.

Answer: 4

Problem 19 Solution

If the number has digits a, b, c, d, e, we know using the divisibility rules for 9 and 11 that

$$a + b + c + d + e \equiv 1 \pmod{9}$$

and

$$e - d + c - b + a \equiv 2 \pmod{11}.$$

We want the largest 5-digit number possible, so we try $a = 9$, $b = 8$, $c = 7$ first. In this case we need

$$d + e \equiv 4 \pmod{9} \text{ and } e - d \equiv 5 \pmod{11}.$$

For mod 11, this implies $(d,e) = (0,5),(1,6),(6,0)$, but none of these work mod 9. Thus we try $a = 9$, $b = 8$, $c = 6$ next. In this case we need

$$d+e \equiv 5 \pmod 9 \text{ and } e-d \equiv 6 \pmod{11}.$$

For mod 11, this implies $(d,e) = (1,7),(5,0)$ and here $d = 5, e = 0$ works. This gives the number 98650.

Answer: 98650

Problem 20 Solution

Without loss of generality assume square $ABCD$ has side length 1. Thus the cube has volume 1, so the pyramid must have a height of 3 so its volume is also 1. As $K > 0$, they overlap as in the diagram below.

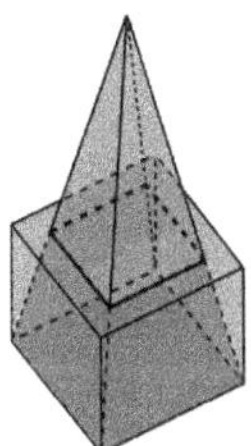

Consider the portion of the pyramid outside the cube. This smaller pyramid is similar to the original with a height of 2. As the height of the smaller pyramid is $\frac{2}{3}$ of the full pyramid, its volume is $\left(\frac{2}{3}\right)^3 = \frac{8}{27}$ of the original. Hence the portion of the full pyramid that is inside the cube is $1 - \frac{8}{27} = \frac{19}{27} \approx 0.703$ of the full pyramid. As the cube and full pyramid have the same volume, $K \approx 70.3$ so rounded to the nearest integer $K = 70$.

Answer: 70

2.9 ZIML June 2018 Junior Varsity

Below are the solutions from the Junior Varsity ZIML Competition held in June 2018.
The problems from the contest are available on p.65.

Problem 1 Solution

Let $r > s$ be the two roots. Using Vieta's formulas we have $r+s=-m$ and $rs=3$. Hence $n=3$. As $r-s=6$ we have

$$\begin{aligned}36 &= (r-s)^2 = r^2-2rs+s^2 = (r+s)^2-4rs\\ \Rightarrow m^2 &= 36+4\cdot 3 = 48.\end{aligned}$$

Hence $m^2+n^2 = 48+3^2 = 57$.

Answer: 57

Problem 2 Solution

The interior angles of a regular polygon with n sides each measure $\dfrac{180^\circ(n-2)}{n}$. Hence the interior angle of the triangle, square, etc., up to the decagon are

$$60^\circ, 90^\circ, 108^\circ, 120^\circ, 128\tfrac{4}{7}^\circ, 135^\circ, 140^\circ, 144^\circ.$$

The sum of these is $925\frac{4}{7}^\circ$, but we want an angle $\leq 180^\circ$, so we first consider the remainder when divided by 360 which is $205\frac{4}{7}^\circ$. This is the reflex angle, so we want

$$360^\circ - 205\tfrac{4}{7}^\circ = 154\tfrac{3}{7}^\circ.$$

Rounded to the nearest integer this is 154 degrees.

Answer: 154

Problem 3 Solution

First arrange the 6 girls. Since they are in a circle there are $6! \div 6 = 120$ ways for them to be arranged.

This creates 6 spots to fill in with the boys. There are $6 \cdot 5 \cdot 4 \cdot 3 = 360$ ways to choose how the boys fill in these spots.

In total there are $120 \cdot 360 = 43200$ arrangements of the students.

Answer: 43200

Problem 4 Solution

Working mod 12 we have

k	0	1	2	3	4	5	6	7	8	9	10	11
$k^2 \pmod{11}$	0	1	4	9	4	1	0	1	4	9	4	1

For every 12 perfect squares, 4 of them will have remainder 1 when divided by 12. As $1000 \div 12 = 83\, R\, 4$, there are $83 \cdot 4 + 1 = 333$ such squares.

Answer: 333

Problem 5 Solution

Since we know the books are automatically grouped by author, we can ignore how many books by each author there are. Then there is one arrangement in order (A, C, H) out of $3! = 6$ total.

Hence $\dfrac{P}{Q} = \dfrac{1}{6}$ so $P + Q = 1 + 6 = 7$.

Answer: 7

Problem 6 Solution

Clearly $x \neq 0$, so we know that

$$2x^2 + 5x - 8 + \frac{5}{x} + \frac{2}{x^2} = 2\left(x^2 + \frac{1}{x^2}\right) + 5\left(x + \frac{1}{x}\right) - 8 = 0.$$

Substituting $y = x + \frac{1}{x}$ we get

$$2(y^2 - 2) + 5y - 8 = 2y^2 + 5y - 12 = (2y - 3)(y + 4) = 0$$

Hence $y = \frac{3}{2}$ or $y = -4$.

Therefore $x + \frac{1}{x} = -4$ or $x + \frac{1}{x} = \frac{3}{2}$. Note however that

$$x + \frac{1}{x} = \frac{3}{2} \Rightarrow 2x^2 - 3x + 2 = 0$$

has no real solutions. Thus $x + \frac{1}{x} = -4$. (To double check, the two real solutions are $x = -2 \pm \sqrt{3}$.)

Answer: -4

Problem 7 Solution

As $AD = DE = EF = FB$, $AF : AB = 3 : 4$ so $[AFC] = \frac{3}{4}[ABC]$ ($\triangle AFC$ has $\frac{3}{4}$ the area of $\triangle ABC$).

Similarly $[AGH] = \frac{1}{3}[AFC]$. Combining we have $[AGH] = \frac{1}{4}[ABC]$.

Using Heron's formula

$[ABC] = \sqrt{12(12-9)(12-8)(12-7)} = \sqrt{12 \cdot 3 \cdot 4 \cdot 5} = 12\sqrt{5}$.

Thus $[AGH] = 12\sqrt{5} \div 4 = 3\sqrt{5}$ and $R + S = 3 + 5 = 8$.

Answer: 8

Problem 8 Solution

$7 \cdot 11 \cdot 13 = 1001$. For any 3-digit number $\overline{abc}$ we have $\overline{abc} \cdot 1001 = \overline{abcabc}$, so our 6-digit number must be of the form $\overline{abcabc}$ to be divisible by 7, 11, and 13.

Using the divisibility rule for 9 we must have $2a+2b+2c$ divisible by 9, so $a+b+c$ must be divisible by 9. To get the smallest 6-digit number possible we have $a=1, b=0, c=8$ giving the number 108108.

Answer: 108108

Problem 9 Solution

Let $\angle AOB = x$ the measure of (minor) arc $\widehat{AB}$. Let $\widehat{CD} = y$, $\widehat{EF} = z$ as in the diagram below

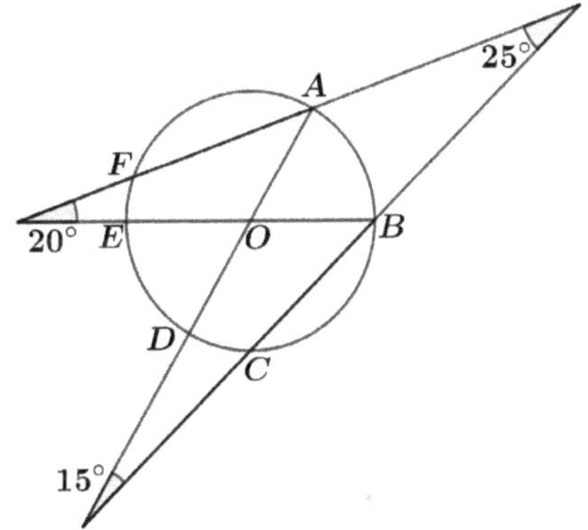

Since O is the center of the circle, we have $\widehat{ED} = x$ as well. Thus

$$20^\circ = \frac{x-z}{2}, 15^\circ = \frac{x-y}{2}, 25^\circ = \frac{x+y+z-x}{2} = \frac{y+z}{2}.$$

Hence $x = y+30^\circ = z+40^\circ$ and $y+z = 50^\circ$ so

$$2x = y+z+70^\circ = 120^\circ.$$

Thus $x = \angle AOB = 60^\circ$.

Answer: 60

Problem 10 Solution

Note that Jaki's brother must be working with a base greater than 5, as she is using the digit 5. If she was working in base 6, the sum would be equal to

$$53_6 + 35_6 = (5 \times 6 + 3) + (3 \times 6 + 5) = 56$$

which is not the same as

$$121_6 = 1 \times 36 + 2 \times 6 + 1 = 49.$$

If she was working in base 7, the sum would be equal to

$$53_7 + 35_7 = (5 \times 7 + 3) + (3 \times 7 + 5) = 64$$

and

$$121_7 = 1 \times 49 + 2 \times 7 + 1 = 64,$$

so Jaki's brother must be working with numbers in base 7.

Answer: 7

Problem 11 Solution

We use complementary counting. Using stars and bars there are $\binom{8+4-1}{8} = \binom{11}{8} = 165$ total ways to buy 8 bottles of soda.

Considering the cases of 0, 1, 2, 3, or 4 bottles of Coke and Sprite, using stars and bars there are respectively

$$\binom{8+2-1}{8} = 9, \binom{6+2-1}{6} = 7, \binom{4+2-1}{4} = 5,$$
$$\binom{2+2-1}{2} = 3, \binom{0+2-1}{0} = 1$$

ways to get the same number of bottles of Sprite and Coke. Hence there are

$$165 - 9 - 7 - 5 - 3 - 1 = 140$$

ways for Norman to buy different numbers of bottles of Sprite and Coke.

Answer: 140

Problem 12 Solution

Multiplying by 2 on both sides we have

$$2x^2+4x-12=2\sqrt{2x^2+4x+3}.$$

Then, making the substitution $y=\sqrt{2x^2+4x+3}$ we get

$$y^2-15=2y \text{ or } y^2-2y-15=(y-5)(y+3)=0.$$

Hence $y=5$ or $y=-3$, which does not work as square roots cannot be negative.

$5=y=\sqrt{2x^2+4x+3}$ so $2x^2+4x-22=x^2+2x-11=0$. Using the quadratic formula we get $x=-1\pm2\sqrt{3}$. Thus the difference between the largest and smallest roots is $4\sqrt{3}$. As $1.7<\sqrt{3}<1.8$ we have $6.8<4\sqrt{3}<7.2$ and thus rounded to the nearest integer, our answer is 7.

Answer: 7

Problem 13 Solution

Note first that in the expansion of $(x^3+2x-1)^3$ there is no term with power x^8 and thus $a=0$. Focusing on the other numbers in our expression (127, 63, 31, etc.) we notice they are all one less than a power of 2. Setting $f(x)=(x^3+2x-1)^3$, we have

$$\begin{aligned}&2^9+a\cdot2^8+b\cdot2^7+\cdots+h\cdot2+i\\&=f(2)\\&=(2^3+2\cdot2-1)^3=11^3=1331\end{aligned}$$

and similarly

$$1^9+a+b+\cdots+h+i=f(1)=2^3=8.$$

Therefore

$$\begin{aligned}1323 &= 11^3 - 2^3\\ &= f(2) - f(1)\\ &= 2^9 - 1^9 + (2^8 - 1^8)a + (2^7 - 1^7)b + \cdots + (2^1 - 1^1)h\\ &= 511 + 255a + 127b + 63c + \cdots + 3g + h\end{aligned}$$

Hence, as $a = 0$, the expression is equal to $1323 - 511 = 812$.

Answer: 812

Problem 14 Solution

Let $x-2$, x, and $x+2$ denote the sides of the triangle. The angle bisector divides the longest side into segments of length $x+2-8.4 = x-6.4$ and 8.4. Using the angle bisector theorem we have

$$\frac{x}{x-2} = \frac{x-6.4}{8.4} \Rightarrow \frac{x}{x-2} = \frac{10x-64}{84}.$$

Cross multiplying we have

$$84x = 10x^2 - 84x + 128 \text{ or } 5x^2 - 84x + 64 = 0.$$

Factoring we have $(5x-4)(x-16) = 0$ so $x = \frac{4}{5}$ or $x = 16$. Only $x = 16$ works, so the triangle has sides 14, 16, 18 with perimeter 48.

Answer: 48

Problem 15 Solution

The prime factorization of 60 is $2^2 \cdot 3 \cdot 5$ so the prime factorization of $60^{24} = 2^{48} \cdot 3^{24} \cdot 5^{24}$.

Therefore perfect square factors will be of the form $2^{2a} \cdot 3^{2b} \cdot 5^{2c}$ for integers a, b, c with $0 \le a \le 24$ and $0 \le b, c \le 12$. Hence there are $25 \cdot 13 \cdot 13 = 4225$ factors of 60^{24} that are perfect squares.

Similarly there are $(16+1)(8+1)(8+1) = 17\cdot 9\cdot 9 = 1377$ factors of 60^{24} that are perfect cubes.

However we are overcounting as some are perfect sixth powers (so both a perfect square and perfect cube). As there are

$$(8+1)(4+1)(4+1) = 9\cdot 5\cdot 5 = 225$$

perfect sixth powers, in total we have

$$4225 - 1377 - 225 = 2623$$

factors in the list that are perfect squares or perfect cubes. One of these is 60^{12} and the others come in $2622 \div 2 = 1311$ pairs that each multiply to 60^{24}. Hence the product of all these is

$$60^{12}\cdot \left(60^{24}\right)^{1311} = 60^{12+24\cdot 1311} = 60^{15744}$$

so $K = 15744$.

Answer: 15744

Problem 16 Solution

If we plot Frankie's temperature and Chloe's temperature as the point (C,F) on a 100×100 square, the region where Frankie's temperature is larger than Chloe's is given below

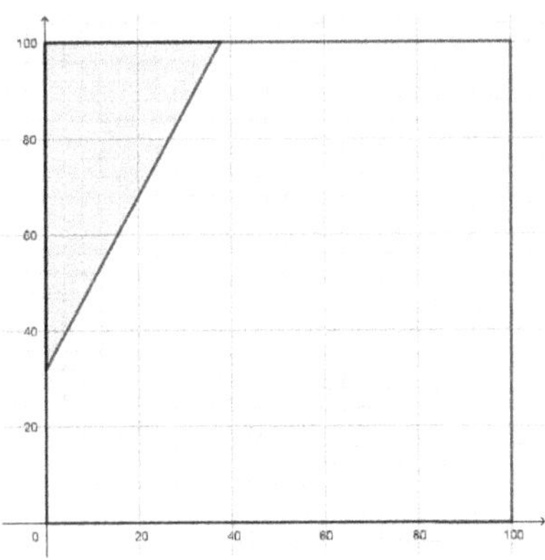

where the hypotenuse of the shaded triangle is the line

$$y = \frac{9}{5}x + 32.$$

Hence the triangle has base $100 - 32 = 68$ and the height is found using the equation $100 = \frac{9}{5}x + 32$. Solving, the height is

$$68 \cdot \frac{5}{9} = \frac{340}{9}.$$

Thus the area of the shaded triangle is

$$\frac{1}{2} \cdot 68 \cdot \frac{340}{9} = \frac{11560}{9} \approx 1284.44.$$

This gives a probability of $\frac{1284.44}{100^2} \approx 12.84\%$. Therefore rounded to the nearest integer, $K = 13$.

Answer: 13

Problem 17 Solution

Rearranging the first equation we have $y = |x| - 4$. Substituting we get, (note $x^2 = |x|^2$),

$$|x|(|x| - 4) + x^2 = 0 \Rightarrow |x|^2 - 4|x| + |x|^2 = 2|x|(|x| - 2) = 0$$

Thus $|x| = 0$ or $|x| = 2$. If $|x| = 0$ we have $x = 0$ and $y = -4$. If $|x| = 2$ we have $x = \pm 2$ and $y = -2$. The maximum value of $x \cdot y$ is thus for the pairs $(-2, -2)$ with $x \cdot y = -2 \cdot -2 = 4$.

Answer: 4

Problem 18 Solution

All the spheres have a radius of 1, so the hexagon has a side length of 2. As a regular hexagon is made up of 6 equilateral triangles, the length between two opposite vertices in the hexagon is $2 \cdot 2 = 4$.

Looking at this side view (through the opposite vertices) gives the diagram below, where r is the radius of the seventh sphere.

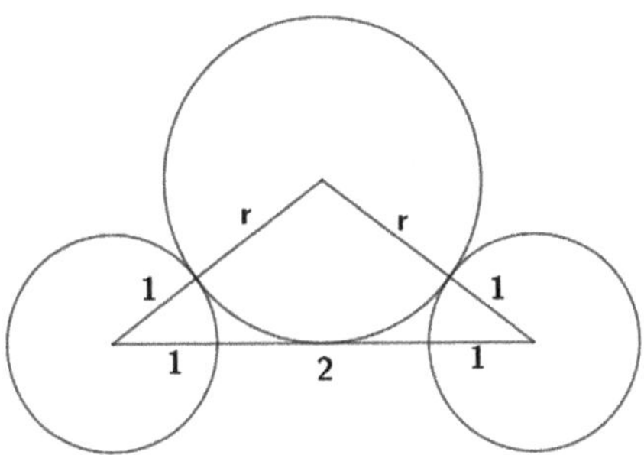

Note by dropping an altitude we divide this triangle into two right triangles with sides 2, r, and $1+r$.

Hence using the Pythagorean theorem we have $2^2+r^2=(1+r)^2$ so $4=1+2r$ and thus $r=\frac{3}{2}=1.5$.

Answer: 1.5

Problem 19 Solution

We know the first 4 blocks give 16 paths in total. If we label Larry's house as $(0,0)$, these paths end at either $(0,4)$, $(1,3)$, $(2,2)$, $(3,1)$, or $(4,0)$. As each step is either N or E, there are respectively

$$\binom{4}{0}=1, \binom{4}{1}=4, \binom{4}{2}=6, \binom{4}{3}=4, \binom{4}{4}=1$$

paths that lead to each of these points. Similar reasoning gives there are, respectively $1,4,6,4,1$ paths leading from

$$(0,4),(1,3),\ldots,(4,0) \text{ back to } (0,0).$$

Therefore

$$1^2+4^2+6^2+4^2+1^2=1+16+36+16+1=70$$

of the 256 paths result in Lost Larry returning home.

Answer: 70

Problem 20 Solution

Working modulo 5, we have $x^2 \equiv 4 \pmod{5}$. Hence either $x \equiv 2 \pmod{5}$ or $x \equiv 3 \pmod{5}$.

If $x = 5n+2$ for an integer n, then

$$\begin{aligned} &(5n+2)^2 - 5y = 4 \\ \Rightarrow\ &25n^2 + 20n + 4 - 5y = 4 \\ \Rightarrow\ &y = 5n^2 + 4n. \end{aligned}$$

Similarly if $x = 5n+3$ for an integer n, then

$$\begin{aligned} &(5n+3)^2 - 5y = 4 \\ \Rightarrow\ &25n^2 + 30n + 9 - 5y = 4 \\ \Rightarrow\ &y = 5n^2 + 6n + 1. \end{aligned}$$

To have $0 \le x \le 100$, we need $0 \le n \le 19$ (for both $x = 5n+2$ and $x = 5n+3$). We also want $0 \le y \le 100$. If $n = 4$,

$$5n^2 + 4n = 96 \text{ and } 5n^2 + 6n + 1 = 105.$$

From this is easily follows that $(5n+2, 5n^2+4n)$ is such a solution for $0 \le n \le 4$ and $(5n+3, 5n^2+6n+1)$ is such a solution for $0 \le n \le 3$. This gives $5+4 = 9$ solutions in total with $0 \le x, y \le 100$.

Answer: 9

3.1 Junior Varsity Topics Covered

Algebra

- Students should be comfortable with ratios, proportions, and their applications to problems involving work and motion, but these problems are not a main focus at this level
- Exponents and Radicals: Laws of Exponents, Simplest Radical Form for Roots
- Factoring Tricks: Sums and differences of squares, cubes, etc., Binomial and multinomial theorem, Completing the Square/Rectangle, etc.
- Solving Equations: Linear Equations, Quadratic Equations, Systems of Equations, Substitutions to rewrite higher degree equations as quadratics, Radicals, Absolute Values
- Quadratics: Graphing and Vertex Form, Maxima and Minima, Quadratic Formula, Discriminant, Vieta's Theorem for sum and product of the roots

- Polynomials: Polynomial Long Division, Remainder and Factor Theorem, Rational Root Theorem, General Vieta's Theorem

Geometry

- As a general rule students should be comfortable using algebraic techniques (linear equations, quadratic equations, systems of equations, etc.) as tools for applying the geometric concepts listed below
- Angles in Parallel Lines (corresponding angles, alternating interior/exterior angles, same-side interior/exterior angles, etc.)
- Analytic Geometry: Equations of Lines, Parabolas, and Circles, Distance Formula, Midpoint Formula, Geometric Interpretation of Slope and Angles
- Triangles: Congruence and Similarity, Pythagorean theorem, Ratios of Sides for triangles with angles of 45, 45, 90 or 30, 60, 90
- Centers in Triangles: Definitions of altitudes, medians, angle bisectors, perpendicular bisectors, Definitions and basic properties of orthocenter, centroid, incenter, circumcenter, Angle Bisector Theorem
- Interior and Exterior Angles of Polygons, including the sum of all these angles, each angle if the polygon is regular, etc.
- Areas and Perimeters of basic shapes such as triangles, rectangles, parallelograms, trapezoids, and circles, Heron's formula and formulas using inradius or circumradius for triangles
- Geometric Reasoning with Areas: Congruent shapes have the same area, Similar triangles have a ratio of areas that is the square of the ratio of their sides, Triangles with the same height have a ratio of their areas equal to the ratio of their bases, etc., Using multiple expressions of area to solve for unknowns

- Circles: Arc Length, Sector Area, Definitions for Tangent Lines and Tangent Circles, Inscribed Angles, Angles formed by intersecting chords, Power of a Point, Ptolemy's Theorem
- Solid Geometry: Surface Area and Volume for Spheres, Prisms, Pyramids, and Cones, Reasoning for more general solids, such as combining the solids listed above or pieces of solids when cut by a plane, etc.

Counting and Probability

- Fundamental Rules: Sum and Product Rules, Permutations and Combinations
- Counting Methods: Complementary counting, Stars and bars (also called sticks and stones, balls and urns, etc.), Grouping objects that must be together, Inserting objects that must be apart into spaces between objects, etc., Principle of Inclusion and Exclusion
- Identities: Symmetry, Pascal's Identity, Hockey Stick Identity, etc. for binomial coefficients, Binomial and Multinomial Theorem, Understanding of these identities using combinatorial proofs
- Sequences: Arithmetic and Geometric Sequences and Series, Finding and understanding patterns and recursive definitions for general sequences
- Probability and Sets: Definitions for event, sample space, complement, intersection, and union, Understanding the use of Venn Diagrams
- Probability: In finite sample spaces as a ratio of the number of outcomes, In geometric sample spaces as a ratio of lengths, areas, or volumes, Axioms of Probability, Independence, Conditional Probability, Law of Total Probability

Number Theory

- Fundamental Definitions: Prime numbers, factors/divisors, multiples, least common multiple (LCM), greatest common factor/divisor (GCF or GCD), perfect squares/cubes/etc.
- Number Bases: Expressing and converting numbers in base 2, 3, 8, 16, etc, Understanding how to perform arithmetic in different bases
- Divisibility Rules for numbers such as 2, 3, 4, 5, 8, 9, 10, 11, and how to combine the rules for numbers such as 6, 22, etc.
- (Unique) Prime Factorization and how to use the prime factorization to find the number of factors, to test whether a number is a perfect square/cube/etc, to find the LCM or GCD.
- Factoring Tricks: Factors come in pairs, perfect squares have an odd number of factors, etc.
- Modular Arithmetic: Connection with remainders and applications such as "find the units digit", General rules for addition, subtraction, multiplication, and division, Extension of divisibility rules to calculating a number modulo 9, 11, etc., Fermat's Little Theorem, Euler's Totient Function and extension to Fermat's Little Theorem

3.2 Glossary of Common Math Terms

Acute Angle An angle less than $90°$.

Altitude of a Triangle A line segment connecting a vertex of a triangle to the opposite side forming a right angle. Also called the height of a triangle.

Angle A figure formed by two rays sharing a common vertex. Often measured in degrees.

Angle Bisector A line dividing an angle into two equal halves.

Arc The curve of a circle connecting two points.

Area The amount of space a region takes up. Often denoted using square brackets: area of $\triangle ABC = [ABC]$.

Arithmetic Sequence A sequence where the difference between one term and the next is constant.

Average See Mean.

Base of a Triangle One side of a triangle, often used when the altitude is drawn from the opposite side to this base.

Binomial Coefficient The symbol $\binom{n}{k} = \frac{n!}{k!(n-k)!}$.

Centroid of a Triangle The intersection of the three medians in a triangle.

Chord A line segment connecting two points on the outside of a circle.

Circle A round shape consisting of points that all have the same distance (called the radius) from the center of the circle.

Circumcenter of a Triangle The intersection of the three perpendicular bisectors in a triangle. Also the center of the circle that circumscribes a triangle.

Circumference The perimeter of a circle.

Circumscribe To draw a shape outside another shape so that the boundaries touch.

Coefficient The number being multiplied by a variable or power of a variable. For example, the coefficient of x^3 in $5x^5+4x^3+2x$ is 4.

Complement In probability, the complement of a set is all elements outside the set.

Composite Number A number that is not prime.

Congruent Two shapes or figures that are exactly the same.

Cube A solid figure formed by 6 congruent squares that all meet at right angles.

Deck of Cards A standard deck of cards has 52 cards. There are 4 suits (clubs, diamonds, hearts, and spades) with each suit having cards of 13 ranks (A (ace), $2,3,\dots,10$, J (jack), Q (queen), and K (king)).

Degree of a Polynomial The highest power of a variable in the polynomial. For example, the degree of $2x^3-5x^6+2$ is 6.

Denominator The bottom number in a fraction.

Diagonal A line segment connecting two vertices of a shape or solid that is not an edge of the shape or solid.

Diameter A chord passing through the center of a circle. The diameter has length that is twice the radius.

Die or Dice A standard die (plural is dice) has 6 sides. Each of the 6 sides has the same chance when the die is rolled.

Digit One of $0,1,2,\ldots,9$ used when writing a number.

Discriminant The expression b^2-4ac for a quadratic equation $ax^2+bx+c=0$.

Distinguishable Objects Objects that are different.

Divisible A number is divisible by another number if there is no remainder when the first number is divided by the second. For example, 35 is divisible by 7.

Divisor A number that evenly divides another number. For example, 6 is a divisor of 48. Also called a factor.

Edge A line segment connecting two vertices on the outside of a shape or solid.

Equally Likely Having the same chance of occurring.

Equiangular Polygon A shape with all equal angles.

Equilateral Polygon A shape with all equal sides.

Equilateral Triangle A regular triangle, one with three equal sides and three equal angles.

Even Number A number divisible by 2.

Exponent The number another number is raised to for powers. For example, in a to the power of b (a^b), the exponent is b.

Face The shape or polygon on the outside of a solid region.

Factor of a Number A number that evenly divides another number. For example, 6 is a factor of 48. Also called a divisor.

Factorial The symbol ! where $n! = n \times (n-1) \times (n-2) \cdots \times 1$.

Fraction An expression of a quotient. For example, $\frac{1}{2}$ or $\frac{9}{7}$.

Function A function is a rule that associates exactly one output with every input. Often described using an equation.

Geometric Sequence A sequence where the ratio between one term and the next is constant.

Greatest Common Divisor/Factor (GCD/GCF) The largest number that is a divisor/factor of two or more numbers.

Incenter of a Triangle The intersection of the three angle bisectors in a triangle. Also the center of a circle that is inscribed inside a triangle.

Indistinguishable Objects Objects that are the same.

Inscribe To draw a shape inside another shape so that the boundaries touch.

Intersecting Lines or curves that cross each other.

Intersection of Two Sets The set of objects that are in both of the two sets. Denoted using $\cap$. For example, $\{2,3\} \cap \{3,4,5\} = \{3\}$.

Isosceles Triangle A triangle with two equal sides and two equal angles.

Least Common Multiple (LCM) The smallest number that is a multiple of two or more numbers.

Mean The sum of the numbers in a list divided by the how many numbers occur in the list. Also called the average.

Median The number in the middle of a list when the list is arranged in increasing order.

Median of a Triangle A line connecting a vertex in a triangle to the midpoint of the opposite side.

Midpoint The point in the middle of a line segment.

Mode The number or numbers occurring most often in a list of numbers.

Multiple A number that is an integer times another number. For example, 72 is a multiple of 8.

Numerator The top number in a fraction.

Obtuse Angle An angle between 90° and 180°.

Odd Number A number not divisible by 2.

Orthocenter of a Triangle The intersection of the three altitudes in a triangle.

Parallel Lines Lines that do not intersect.

Perfect Cube A number that is another number cubed. For example, $64 = 4^3$ is a perfect cube.

Perfect Square A number that is another number squared. For example, $64 = 8^2$ is a perfect square.

Perimeter The length/distance around the outside of a shape.

Perpendicular Bisector A line perpendicular to and passing through the midpoint of a line segment.

Pi (π) A number used often in geometry. $\pi = 3.1415926\ldots \approx 3.14 \approx \frac{22}{7}$.

Polygon A shape formed by connected line segments.

Polynomial A function that is made of adding multiples of powers of a variable. For example, $f(x) = x^4 + 3x^2 + 2x - 3$.

Prime Factorization The expression of a number as the product of all its prime factors. For example, 24 has prime factorization $2 \times 2 \times 2 \times 3 = 2^3 \times 3$.

Prime Number A number whose only factors are one and itself.

Proportional Ratios Ratios that have equal values when expressed in fraction form. For example, $2:3$ is proportional to $8:12$.

Quadratic A polynomial with degree 2. Often written in the form $ax^2 + bx + c$.

Quadrilateral A shape with four sides.

Quotient The integer quantity when dividing one number by another. For example, the quotient of $38 \div 5$ is 7 as $38 = 7 \times 5 + 3$.

Radius of a Circle The distance from the center of the circle to any point on the outside of the circle.

Randomly Chosen for a group of objects. Unless specified, the chance of choosing each object is the same as any other object.

Rank of a Card See Deck of Cards.

Ratio A relation depicting the relation between two quantities. For example $2:3$ or $\frac{2}{3}$ denotes that for every 3 of the second quantity there are 2 of the first quantity.

Rational Number A number that can be written as a fraction.

Reciprocal One divided by the number. For example, the reciprocal of 7 is $\frac{1}{7}$.

Rectangle A quadrilateral with four right angles (an equiangular quadrilateral).

Regular Polygon A polygon with all equal sides and all equal angles (equilateral and equiangular).

Remainder The quantity left over when one integer is divided by another. For example, the remainder of $38 \div 5$ is 3 as $38 = 7 \times 5 + 3$.

Rhombus A quadrilateral with four equal sides (an equilateral quadrilateral).

Right Angle A 90° angle.

Right Triangle A triangle containing a right angle.

Root of a Function A value of x such that the function evaluates to zero. For example, $x = 2$ is a root of the function $f(x) = x^2 - 4$.

Sample Space In probability, the sample space is the set of all outcomes for an experiment.

Scalene Triangle A triangle with three unequal sides and three unequal angles.

Sector The region formed by an arc and the two radii connecting the ends of the arc to the center of the circle.

Sequence An ordered list of numbers.

Set An unordered collection or group of objects without repeated elements. Denoted using curly brackets. For example, $\{1,2,3,4\}$ is the set containing the integers $1,\ldots,4$.

Similar Shapes or solids that have the same angles and sides that share a common ratio.

Simplest Radical Form An expression containing a radical such that the number inside the radical is an integer that has no perfect squares.

Sphere A round solid consisting of points that all have the same distance (called the radius) from the center of the sphere.

Square A shape with four equal sides and four equal angles (a regular quadrilateral).

Subset A set of objects that is contained inside a larger set of objects. Denoted using $\subseteq$. For example $\{2,3\} \subseteq \{1,2,3,4\}$.

Suit of a Card See Deck of Cards.

Surface Area The total area of all the faces of a solid.

Tangent Line A line touching a shape or curve at exactly one point.

Trapezoid A quadrilateral with one pair of parallel sides.

Triangle A shape with three sides.

Union of Two Sets The set of objects that are in one or both of the two sets. Denoted using $\cup$. For example, $\{2,3\} \cup \{3,4,5\} = \{2,3,4,5\}$.

Venn Diagram A diagram with circles used to understand the relationship between overlapping sets.

Vertex The intersection of line segments, especially the intersection of sides or edges in a shape or solid.

Volume The amount of space a solid region takes up.

With Replacement When choosing objects with replacement, a chosen object is returned to the others allowing it to be chosen more than once.

3.3 ZIML Answers

ZIML October 2017 Junior Varsity

Problem 1: 43200

Problem 2: 39

Problem 3: 7

Problem 4: -5

Problem 5: 98721

Problem 6: 30

Problem 7: 150

Problem 8: 40

Problem 9: 870

Problem 10: 9

Problem 11: 48

Problem 12: 25

Problem 13: 48

Problem 14: 34

Problem 15: -7

Problem 16: 40320

Problem 17: 63

Problem 18: 15

Problem 19: 3

Problem 20: 24

ZIML November 2017 Junior Varsity

Problem 1:	106	Problem 11:	45
Problem 2:	8	Problem 12:	60
Problem 3:	−26	Problem 13:	16
Problem 4:	240	Problem 14:	28
Problem 5:	432	Problem 15:	127
Problem 6:	13	Problem 16:	6961
Problem 7:	24068	Problem 17:	3
Problem 8:	7	Problem 18:	3
Problem 9:	168	Problem 19:	9
Problem 10:	0.67	Problem 20:	21

ZIML December 2017 Junior Varsity

Problem 1:	12	Problem 11:	5.75
Problem 2:	77148	Problem 12:	120
Problem 3:	1944	Problem 13:	-5
Problem 4:	5	Problem 14:	1
Problem 5:	491	Problem 15:	2187
Problem 6:	310	Problem 16:	15
Problem 7:	-10	Problem 17:	44
Problem 8:	41	Problem 18:	10
Problem 9:	28	Problem 19:	45
Problem 10:	34	Problem 20:	150

ZIML January 2018 Junior Varsity

Problem 1:	53	Problem 11:	33
Problem 2:	37	Problem 12:	48
Problem 3:	89	Problem 13:	6
Problem 4:	36	Problem 14:	349
Problem 5:	−9	Problem 15:	26
Problem 6:	3	Problem 16:	0
Problem 7:	192	Problem 17:	72576
Problem 8:	−0.4	Problem 18:	7
Problem 9:	8.3	Problem 19:	7
Problem 10:	217	Problem 20:	16

ZIML February 2018 Junior Varsity

Problem 1: 71

Problem 2: 21

Problem 3: 34

Problem 4: 0

Problem 5: -9

Problem 6: 800

Problem 7: 48

Problem 8: 3306

Problem 9: 8

Problem 10: 45

Problem 11: 23

Problem 12: 402

Problem 13: -1.5

Problem 14: 70

Problem 15: 1.4

Problem 16: 476

Problem 17: -0.33

Problem 18: 126

Problem 19: -64

Problem 20: 2304

ZIML March 2018 Junior Varsity

Problem 1:	78965	Problem 11:	60
Problem 2:	8.3	Problem 12:	44
Problem 3:	126	Problem 13:	54
Problem 4:	-6.2	Problem 14:	8
Problem 5:	210	Problem 15:	4
Problem 6:	6	Problem 16:	6
Problem 7:	-3	Problem 17:	210
Problem 8:	73	Problem 18:	4
Problem 9:	344	Problem 19:	3
Problem 10:	11	Problem 20:	45

ZIML April 2018 Junior Varsity

Problem 1: 43

Problem 2: 0

Problem 3: 510

Problem 4: 45

Problem 5: 266670

Problem 6: 2015

Problem 7: 12

Problem 8: 106

Problem 9: 1147

Problem 10: 5

Problem 11: 34

Problem 12: 4

Problem 13: 25

Problem 14: 1.7

Problem 15: 34

Problem 16: −1

Problem 17: 23

Problem 18: 0

Problem 19: 39

Problem 20: 10

ZIML May 2018 Junior Varsity

Problem 1:	6	Problem 11:	20
Problem 2:	59	Problem 12:	50
Problem 3:	22.5	Problem 13:	10
Problem 4:	1	Problem 14:	2147
Problem 5:	330	Problem 15:	4.5
Problem 6:	33	Problem 16:	720
Problem 7:	49	Problem 17:	58
Problem 8:	24	Problem 18:	4
Problem 9:	-1	Problem 19:	98650
Problem 10:	36	Problem 20:	70

ZIML June 2018 Junior Varsity

Problem 1:	57	Problem 11:	140
Problem 2:	154	Problem 12:	7
Problem 3:	43200	Problem 13:	812
Problem 4:	333	Problem 14:	48
Problem 5:	7	Problem 15:	15744
Problem 6:	-4	Problem 16:	13
Problem 7:	8	Problem 17:	4
Problem 8:	108108	Problem 18:	1.5
Problem 9:	60	Problem 19:	70
Problem 10:	7	Problem 20:	9

www.ingramcontent.com/pod-product-compliance
Lightning Source LLC
Chambersburg PA
CBHW072352200326
41519CB00015B/3744

9781944863302